GeoSPATIAL MATTERS

GeoSPATIAL MATTERS

Exploring the Implications of a Digital Earth

An edited volume of *GeoWorld/GIS World* Columns

Daniel Z. Sui and
Matt Ball, Co-Editors

GeoTec Media
A division of M2MEDIA360

Geospatial Matters: Exploring the Implications of a Digital Earth

Daniel Z. Sui and Matt Ball, Co-Editors

2006 GeoTec Media
A division of M2MEDIA360, San Francisco, California, USA

Publisher: Jo Treadwell
Layout and Design: Kathleen Sage
Cover Image: The oblique montage of the Longmont and Boulder, Colo., area was created by Rick Lovel. Data was assembled and prepared in ArcMap and Global Mapper, then seamlessly imported to Visual Nature Studio (VNS). Terrain is the USGS National Elevation Dataset, draped with mosaiced 1:24000 Digital Raster Graphics (topo). A vector shapefile of water bodies was added on top of the DRG mosaic. Terrain textures (including snow) are applied and dynamically driven in VNS by the slope, aspect, and elevation parameters inherent in the underlying terrain data. For a touch of realism and depth, the image was finished off with a volumetric atmosphere. (More of Lovel's work can be seen in his gallery on the 3D Nature Web site *www.3DNature.com*.)
Production Manager: Betty Wu
Production Assistant: Sue Gillerlain
Copy Editor: Todd Danielson

Library of Congress Control Number: 2006925409

Geospatial Matters: Exploring the Implications of a Digital Earth / Daniel Z. Sui and Matt Ball, co-editors
283 p. 19x23 cm.
Includes bibliographical references.
ISBN 0-9748613-0-8

Published by GeoTec Media, a division of M2MEDIA360,
760 Market Street, Suite 432, San Francisco, CA 94102, USA
www.geoplace.com/books

Contents

Section 2. GIS, History, and Politics

Section 3. GIS, Spatial Analysis, and the Art of Mapping

Section 4. GIS and Hot-Button Issues

Section 5. GIS, Ethics, and Education

Section 6. GIS, Fieldwork and Ubiquitous Computing

Section 7. Digital Earth and the Impact of GIS

Afterword

Foreword

We are in the process of a major shift in how we see and manage our world. Our environment, which was largely controlled by natural processes in the past, is becoming increasingly affected by human activities. The role of geography in society has been one of recording, documenting, and describing the world and all that's in it. It is focused on measuring and classifying places and modeling geographic processes of change.

With advances in computing, geography has been evolving toward more quantitative approaches incorporating advanced measurement and digital processing techniques. At the same time, we see the emergence of widespread use of GIS to organize, manage, and provide access to a virtual digital world of geographic information. GIS technology is making it possible to better apply geographic knowledge to the world's problems and will increasingly be important as a framework for global evolution.

Fundamentally, at the heart of GIS, is a whole new way for human beings to abstract their world. Since the beginning of recorded history, humans have used various abstractions to record, communicate, and share their experiences. These include various kinds of languages and graphic expressions, such as hieroglyphics, written languages, music, mathematics, drawings, photographs, paintings, and so on. These abstractions represent our collective understanding, our common knowledge, and in many ways describe our civilization. GIS represents a kind of new abstraction, a new framework for organizing geographic knowledge and a new language.

GIS is used to transactionally maintain the status and history of geography using geodatabase technology. There are many types of transactions that update geodatabases including new measurements (i.e., satellite images, land surveys, etc.), changes (i.e., administrative boundary changes, land subdivisions, etc.), and events that alter the landscape (i.e., construction, disasters, migration, etc.).

GIS is growing rapidly with numerous applications in many industries. Application examples give us the evidence of its overwhelming success in government, education, and business. Almost every sector from national security, conservation, real estate site selection, and management of natural resources has benefited. The generic causes driving this success include contributions to efficiency, better decision making, communication, and science. We are also seeing GIS as an instrument (like a microscope) for education and learning. Finally, GIS provides a collaborative problem solving framework wherein multiple disciplines, organizations, and interests can work together in geographic planning exercises and decision making. These include land use planning, emergency crisis response, and environmental conservation.

As GIS continues to expand in its acceptance and application, we are seeing growing interest in developing GIS as a societal infrastructure for providing geographic knowledge access to everyone. This vision is possible because of the rapid evolution of GIS technology on the Internet with geographic information servers and interoperability standards.

It has been suggested that fast computing, the ability to store large amounts of geographic information, and very fast communication networks together with GIS software advances, will ultimately lead to a global nervous system that serves up our collective geographic knowledge and makes it accessible for many applications ranging from citizen awareness to professional decision making and science. This vision is being supported by the rapid progress in underlying technology and also the increasing need for better geographic information to support nearly all human activities.

Is this vision reasonable and practical? Yes, the natural evolution of technology and our social institutions' acceptance of this infrastructure are evolving rapidly. However, there remain a number of issues that will need to be addressed before a global GIS infrastructure can be created, such as technical interoperability, data content standards, policies for information sharing, GIS organization, data management, financing and sustainability of GIS infrastructure across organizations, and developing an experienced workforce. All of these issues can and will be solved. It will require ongoing collaboration and sharing of information about best practices.

One of the keys to realizing the full potential of GIS is maintaining an open dialogue among GIS professionals. The columnists who have contributed to GIS World and *GeoWorld* since 1990 are important participants in this dialogue. They ask probing questions that make GIS practitioners stop and think—questions about the direction of GIS technology, GIS as a profession, and the role of GIS in our society. This comprehensive collection of columns enables GIS users to revisit some of the pressing questions asked over the last 16 years. It also provides some historical context for students and professionals who are new to GIS.

Geographic knowledge is important in our world. People and organizations are beginning to understand the power of geographic information and the value that GIS provides in nearly everything they do. They increasingly want access and the ability to consider all the factors. They want to get the geographic advantage. GIS promises to provide our societies with more clear and quantitative information that can be used to understand these global changes and address problems. It helps us to see, plan, and manage our world.

We are still only at the beginning. We are just now beginning to connect our individual GIS efforts into networks of open and integrated systems. Creating a global GIS infrastructure will require participation by many individuals and organizations acting on their own and acting together. Perhaps you are a GIS user who is already taking the steps necessary to make this vision a reality; if so, I applaud your efforts. For those of you who are new to GIS, I hope that reading this book will encourage you to join GIS users worldwide who are using geography to improve our world.

Jack Dangermond, President, ESRI

Preface

GeoWorld magazine has been privileged through the years to have a thoughtful group of columnists that have explored the implications, rather than simply the application, of geospatial technology.

Within these pages, leading geospatial pundits Jerry Dobson, Nigel Waters, Daniel Sui and Jeremy Crampton tackle such questions as "Can GIS and Geopolitics Prevent War?" and "Is GIS a Privacy Threat?" Their writings contribute greatly to our understanding of geospatial technology in historical, intellectual, scientific, political and social contexts.

Longstanding Vision

Geospatial technology industry pioneers have long held a broad vision about the importance of location to catalog information. The digital revolution has come a long way in the last decade, and many visions that were once far-fetched have become reality.

Former Vice President Al Gore laid out his vision for a "Digital Earth" in January 1998. The vision discussed the ability to "virtually fly through space down through progressively higher-resolution datasets to hover above any point on Earth's surface—and then display information relevant to that location from an infinite number of sources."

The realization of this vision is upon us now with the advent of Geographic Exploration Systems, such as NASA's World Wind, Google Earth and ArcGIS Explorer. These systems broaden the scope of geospatial technology use by providing a simpler interface for geospatial data exploration and a compelling 3-D environment. They build on decades of geographic information system (GIS) development by taking advantage of the geospatial data created by GIS practitioners.

Geographic Exploration Systems enhance the GIS toolbox by providing an open platform to bring together different geospatial data, images and services. The open platform allows for the "mashing" of available geospatial information for greater clarity and understanding of specific places.

Social Issues Persist

As we sit on the cusp of realizing the Digital Earth, the social implications of geospatial technology and the role it plays in science are of greater interest now than ever.

Geospatial technology has a significant role to play in such noble pursuits as safeguarding public health, combating poverty, fighting crime and terrorism, and mitigating manmade and natural disasters. However, the technology is "surveillant" in nature (Waters, 1994), and safeguards are warranted to ensure that practitioners don't manipulate the tools' power to harm others.

Practitioners and scholars must not be wooed by the power without thinking of the purpose. This book achieves a valuable awareness of the "dark side" of misused technology as well as the pitfalls of relying too much on a system when bad data or methodology may be clouding decisions.

Providing Perspective

Many of the contributions in this book predate the Internet's explosive growth, yet, as you read these columns, you'll realize that the questions, ideas and perspectives are no less relevant today.

The breadth of contributions, dating back to 1990 at the earliest, provide a valuable measure of the technology's evolution as well as an awareness that the same challenges and issues have been around for some time.

This historical view helps educate new users as well as remind early adopters about how far we have come.

Roots

GeoWorld, which began as *GIS World* in 1988, was founded by H. Dennison (Denny) Parker as a newsletter that was published out of his home while maintaining his "day job" at Colorado State University. The publication filled a much-needed niche and became rapidly successful, because no other publication focused on what was then an emerging technology.

The company quickly grew and was spurred by Parker's drive to be the "world's leading geographic information systems publication." This drive for excellence was rewarded by readers and advertisers, spurring additional international magazines, directories, conferences, workshops, books and the GeoPlace.com Web portal.

"I always believed that *GIS World* should provide products 'by GISers for GISers'," noted Parker in a 1996 interview. This focus on the passionate user has served the magazine and industry well for almost 20 years.

Audience

When contemplating the idea of this book, which my co-editor Daniel Sui spearheaded, we realized that no existing book takes a broad look at the role of GIS and its contributions to science and society. These columns take on the tough questions, and they do so using accessible language that sparks thought and, hopefully, discussion.

Anyone interested in the use of geospatial technology will benefit from this book. The content is easily accessible to all levels of GIS practitioners as well as the general public. It's a strong classroom resource to augment more technical texts. It also will be of interest to practitioners hoping to gain a broader perspective.

The intellectual curiosity within these pages is infectious, and we hope that this book sparks or renews your passion for the exploration and insight that can only be achieved through the application of geospatial technology.

Matt Ball
Editor, Associate Publisher
GeoWorld Magazine

Contributors

JEREMY CRAMPTON is associate professor of geography at Georgia State University. Originally from the United Kingdom, he obtained his Ph.D. from Penn State in 1994. After a brief stint back in the United Kingdom at the University of Portsmouth, he returned to the United States to pursue a career in academia. His main interest is the politics of mapping and GIS. He began his column in *GeoWorld* on "Critical GIS" in January 2003.

Crampton's previous book, *The Political Mapping of Cyberspace* (Chicago, 2004), argues that questions of space are political. Because mapping, cartography and GIS are important ways to gain knowledge about the world, they also have a politics. The book focuses on cyberspace, arguing that virtual spaces have an interesting set of political questions. The digital divide, for example (the gap between those with access to the information economy and those without), has a distinct geography, so cyberspace isn't evenly spread.

More recently, he has explored the use of maps in the Versailles Peace Conference after World War I and how they were used to make political decisions about post-war Europe.

E-mail: *jcrampton@gsu.edu*

JEROME E. (JERRY) DOBSON is president of the American Geographical Society and professor of geography at the University of Kansas. He holds a Ph.D. in geography from the University of Tennessee (1975), M.S. and B.A. in geography from the University of Georgia (1972, 1967), and A.A. from Reinhardt College (1965). He's a fellow of the Royal Geographical Society and corresponding individual member of the International Geographical Union.

Dobson previously served as a distinguished research and development staff member at Oak Ridge National Laboratory, president of the University Consortium for Geographic Information Science, contributing editor of *GeoWorld* magazine, and U.S. delegate and expert to the International Standards Organization.

Dobson currently leads a project funded by the Geneva International Centre for Humanitarian Demining to field test a GIS/GPS-based system for mapping minefields without walking on them as well as develop and promulgate a new world standard for cartographic symbolization of landmines, mine fields and mine actions. He also leads a project funded by the Department of Defense to collect GIS data and conduct geographic research in Mexico.

Dobson's principal contributions include the paradigm of automated geography, his instrumental role in originating the National Center for Geographic Information and Analysis, his leadership of the LandScan Global Population Project, and his leadership to advance remote-sensing methods for large-area change analysis as part of NOAA's Costal Change Analysis Program. LandScan recently gained widespread acclaim as the only feasible means of estimating populations impacted by the 2004 tsunami in Southeast Asia.

Employing GIS and automated geographic methods, Dobson has proposed new evidence and theory regarding the mechanisms responsible for lake acidification as well as new evidence and theory regarding continental drift and plate tectonics.

E-mail: *dobson@ku.edu*

DANIEL Z. SUI is professor of geography and holder of the Reta A. Haynes endowed chair in the College of Geosciences at Texas A&M University. Sui also serves as assistant vice president for research and director for Geospatial Information Science and Technology at Texas A&M. Sui held a visiting professor position at the University of California—Santa Barbara in 2001 and served as a senior research scientist at the United Nations Food and Agricultural Organization headquarters in Rome in 1999. Sui is the current North America editor for the *International Journal of Computers, Environment and Urban Systems*. He succeeded Jerry Dobson as the "'G' in GIS" columnist in December 2004.

Sui holds a B.S. (1986) and M.S. (1989) from Peking University as well as a Ph.D. from the University of Georgia (1993). His professional research interests include fundamental theoretical issues in geospatial information science and technology, urban sustainable development, GIS applications in public health, and environmental policy issues in the information age. He is author/co-author of four books and more than 100 outstanding scientific publications. He was the Big 12 Faculty Fellow in 1997, AT&T Faculty Fellow in 2000 and ORISE Fellow in 2004. Sui is also the recipient of two outstanding teaching awards at Texas A&M University. For more detailed information, please visit *geog.tamu.edu/~sui*

E-mail: *sui@geog.tamu.edu*

NIGEL WATERS received a first-class honours degree from the University of Cambridge in 1972. He then had the good fortune to study for his Masters and Ph.D. degrees at the University of Western Ontario under the supervision of Mike Goodchild.

Waters began teaching GIS in 1975 at the University of Calgary where, in 1990, he was promoted to full professor. He began writing his "Edge Nodes" column in 1989. In 1999, he became the founding director of the Masters in GIS Program at the University of Calgary. This program has been named one of the five Centers of Excellence worldwide for the use and application of the object-oriented eCognition remote-sensing software.

Waters' research interests include GIS-T, statistical and spatial methodologies, and Web-based GIS mapping. He's the technical director of the Mapping the Media in the Americas Project (*www.mediamap.info*), a collaboration with the Carter Center in Atlanta and Canadian Foundation for the Americas in Ottawa. The research's primary objective is to map the location of the media and demonstrate its influence on the electoral process in Canada and in 11 Latin American countries.

In 2005, Waters co-authored a monograph on the use of Web-based GIS for public participation in transportation planning. His GIS research and publications also reflect his interest in working with First Nations and indigenous peoples. He has written numerous journal articles and book chapters on GIS-T, and statistical and environmental applications of GIS. He's also the president of GIYes Solutions.

E-mail: *nwaters@ucalgary.ca*

Section 1.
Geography, GIS, and Science

Introduction
By Jerry Dobson

Geography is an essential science. For 2,500 years it has led the way toward new understandings of people, places, and processes. Geographer and cartographer Abraham Ortellius, for instance, first proposed continental drift in 1596. Biogeographer Alfred Russell Wallace discovered evolution on his own and might be credited as the discoverer under today's rules of precedence because he published before Charles Darwin. Geographers, starting with Bill Denevan in 1961, forged revolutionary new understandings of the Americas before Columbus that only now are reaching the public through Charles Mann's *1491*.

Today, geography has a new tool that makes it vastly more powerful than ever. The microscope and telescope revolutionized science in their day, and the same can be expected of this new macroscope called GIS. Already, GIS is enticing other disciplines to adopt methods and approaches–spatial analysis, place-based analysis, and scientific integration–that have been mainstays of geography for centuries. Geography is about space, time, entities, and processes, and that's how the real world operates. Yet, many specialized disciplines have rejected or ignored spatial logic, and many of their most popular theories have never been challenged by adequate spatial testing.

As GIS spreads within and beyond the geographic community, more and more tests will be performed, and cherished theories will yield to earth-changing insights. But, change of such magnitude will not come easy. Geographers, GIS practitioners, and GIS-armed renegades among the specialized disciplines themselves will face ridicule, scorn, ostracism, and worse merely for stating what the spatial evidence conspicuously reveals. Only by joining the fray, however, and occasionally drawing blood will we establish ourselves as a respected force in the upper echelons of science, science policy, and public policy influenced by science.

Collectively, the writings in this section ask (and sometimes answer) how geography, GIS, and GIScience fit in the broad framework of science. My own contributions cast geography as uniquely spatial, integrative, and place-based and define GIS as a digital representation of the Earth, structured to support analysis. Thus, GIS is both a computerized model of the real world and the instrument by which geographers and many others view, study, and understand its workings. GIScience focuses on advancing the already wondrous instrument, which automated geography employs it to understand earth features and processes. Nigel Waters places such matters in the context of recent intellectual movements (technologies of knowing, consilience), and asks how GIS can save a discipline that, despite its past glories and present accomplishments, has been under siege since World War II. Daniel Sui tackles recent pronouncements about the death and resurgence of geography as a force in economics, and draws enlightening analogies between GIS and the arts.

What in the World is a Geographer?

By Jerry Dobson

From 1970 to 1976, I served in the U.S. Army National Guard as I completed my Ph.D. in geography at the University of Tennessee. One Saturday afternoon, during field maneuvers, our mechanized infantry company took a break on a grassy knoll overlooking the Tennessee River. Relaxing in the sunshine, I gradually realized that an argument was raging around me. Guardsmen of all enlisted ranks were vehemently debating which city was the capital of Washington state. Some said it was Seattle; others said it was Spokane.

I listened a while, then interjected that the answer was Olympia. My opinion was ignored. The debate went on, each soldier bolstering his claim with recollections from fifth grade geography or the travel experience of a friend or relative. Twice more I offered my opinion. Each time it was dismissed without serious consideration. I gave up.

On Sunday morning we reassembled in the field. One of my fellow sergeants, a strong protagonist, had consulted an atlas at home the night before. He walked up, stared at me with fire in his eyes, and said, "What are you, some kind of damned doctor of geography or something?"

A similar experience happened not long ago when I served as a juror in Loudon County, Tenn. The panel before mine was questioned long and hard. One juror was asked to step down, and I was added to the already-seated panel. The defense lawyer could have asked me all of the questions he asked the other panel members. Instead, he approached me with great seriousness and said, "Mr. Dobson, I have just one question. For my own curiosity, what does a geographer do at Oak Ridge National Laboratory?"

Public Perception

These stories illustrate perceptions that plague geographers throughout their careers. Most people naturally adopt definitions of geography based on whatever courses they had in school. For those never exposed to the discipline beyond elementary school, the "states-and-capitals" view prevails. College graduates, most of whom took a single introductory course, usually consider geography a social science if their course was cultural or confuse geography with geology if their course was physical. Among the public, these conceptions are vastly more common than the definitions that geographers themselves would offer.

The public hardly can be blamed for its confusion, because the science of geography is rarely explained to anyone except graduate students in university geography departments. Most geographers agree the discipline is defined not by its content, but by its spatial approach. Yet, ironically, the undergraduate curriculum focuses on content and techniques to the virtual exclusion of fundamental methods. In fact, it is possible to get a bachelor's degree in most geography departments without taking a single course in geographic methodology or history of geographic thought. Small wonder most dictionaries meagerly define the discipline as a descriptive science of earth features!

Most people can pass blissfully through an entire lifetime without knowing or caring what the true definition is, unless they are faced with a contradiction that stuns them and spurs curiosity. To many the notion of "a geographer at Oak Ridge National Laboratory" creates such a contradiction. Of what possible use could a "states-and-capitals" expert be at a high-technology research and development center? Even the college (cultural vs. physical) definitions are difficult to reconcile with the limited views many people have of the national laboratories. Whatever their views of geography and institutions, many upon hearing my occupation are compelled to ask, "What is a geographer?"

The Geographic Approach

In the *1991-92 International GIS Sourcebook*, David Mark and Holly Dickinson discussed a strong bond between GIS and each of William Pattison's four traditions of geography (Pattison, 1964). These grand traditions—spatial analysis, area studies, man-land (human ecology) studies, and earth science—serve as a good synopsis of what geographers do. This heritage may sound diverse, but there is a common thread: space. The geographic approach, across all four traditions, can be characterized as investigations of (1) the distribution of an individual phenomenon through space; (2) the interrelationships among phenomena that cause their distributions to vary through space and time; (3) the interrelationships that occur among phenomena because of their coincident location at a specific place or places; (4) the physical and cultural character of a specific region or landscape that exists because of the area's unique combination of spatial phenomena; (5) the theory, methods, models, and techniques required to conduct geographic research and analysis as described in items 1 through 4.

In short, professional geographers are scientists and analysts who devote their careers to the pursuit of spatial logic—a form of reasoning in which morphology, spatial distribution, spatial association and temporal association are viewed as primary evidence of collective earth processes. Undoubtedly geography's emphasis on space and spatial relationships accounts for the bond with GIS, both in name and function.

Who are the Good, the Bad, and the Ugly in Geography?

The reaction to "a geographer at Oak Ridge National Laboratory" has caused me to think about geographers in unexpected occupations. In this vein, I offer a trivia quiz:
1. Who is the richest and most famous geographer?
2. Who is the most compassionate geographer?
3. Who is the most militant geographer?
4. Who is the most diabolical geographer?
5. Who is the most philosophical geographer?

Answers:
1. *Richest and Most Famous*: Michael Jordan. Yes, "Air Jordan" has a degree in geography from the University of North Carolina. You may never witness a more powerful display of raw spatial intelligence than the feats he performs in a 3-D space while spinning in the air, reckoning independent trajectories of the basketball and several other players on both teams, and keeping perfect registration of the goal's precise geocoordinates. (Partial credit for those who guess Environmental Systems Research Institute (ESRI) president Jack Dangermond, who ranks high in one measure, but is slightly less famous than the slightly taller Jordan. There is no truth to rumors of a Nike contract involving promotion of a new "Ground Danger" self-navigating walking shoe.)

2. *Most Compassionate*: Mother Teresa. The recipient of the 1979 Nobel Peace Prize first went to India to teach geography, later requested permission to work with the poor of Calcutta, and became a nurse. (Partial credit for those who guess George Demko. As director of the Geography and Regional Science Program of the National Science Foundation, geographer of the State Department and now Director of the Rockefeller Center at Dartmouth College, he has promoted scientific advancement, global peace, and environmental stewardship, but he can be a little gruff at times.)

3. *Most Militant*: General Augusto Pinochet Ugarte. This former president of Chile wrote a textbook on military geography. (Partial credit to East Tennessee State University's Herr Professor Dr. Robert Peplies, who guessed the answer is Karl Haushofer, a Bavarian general and professor of geopolitics at the University of Munich whose writings helped provide a pseudoscientific foundation for Adolf Hitler's "Lebensraum" campaign of expansion. Rudolph Hess was one of Haushofer's students and may have introduced the ideas to Hitler.)

4. *Most Diabolical*: Gerard J. Schaefer, Jr. This serial killer chose his victims based on his obsession with geography. He was convicted in 1973 after killing a teenage girl named Georgia Jessup (there is a town named Jessup, Ga.) and her friend whose last name was Place (a fundamental concept in geography). Evidence in his

home suggested he had killed as many as six other women with ethnic names (eg., Carmen, Nina, and Marguerita). In unfortunate juxtaposition, a news article of the day (*Time*, 1973) says he was found to have "impairment in reality testing" and "got a degree in geography" at Florida Atlantic University in 1968.

5. *Most Philosophical*: Immanuel Kant. This German scholar is known throughout the world as a great philosopher, but he liked to teach physical geography. Perhaps exploring the global diversity of physical environment and human culture inspired his famous philosophy, "Alles ist relativ (Everything is relative)."

(This column originally appeared in the July 1993 issue of GIS World.*)*

The Two Faces of Geography

By Jerry Dobson

Geography is Dead! Test after test has shown that U.S. high school and college students can't locate anything on a world map. Canada is a mirror image of the Gulf of Mexico. France is lost forever. Geography is no longer appreciated or understood by our children. Doonesbury knows. Dan Rather reports. The evidence is clear. The litany is familiar.

Geography is Booming!

In the world of high technology, geography is on a roll. The hottest item at the tip of the information revolution is the geographic information system, or GIS. Estimates of GIS markets for the United States range as high as $3 billion for 1990. [Update: The worldwide GIS market reached an estimated $7 billion in 2000 and $30 billion in 2005.] GIS specialists are in such demand that most university geography departments cannot compete in the hiring competition. [Update: in 2005 the U.S. Department of Labor listed "geospatial technologies" as one of three emerging employment fields with annual employment growth rates projected at 10 to 20 percent.] The National Science Foundation invested $12 million to establish a National Center for Geographic Information and Analysis (1988-1993). [Update: This grant was later reduced to $ 5.5 million, as explained in a subsequent column, "Remembering The Long Hard Battle for NCGIA" (Dobson 1992). Later, fortunately, the Center won additional NSF funding for the Varenius Project, and the University of California at Santa Barbara was awarded substantial funding for the Alexandria Digital Library Project.] The future of scientific and technical geography looks brighter than at any other time since the Middle Ages.

How can such a contradiction exist between the dismal public perception of geography and the excitement that reigns on the high-tech stage? The answer lies largely in definition. To the professional geographer, the field includes far more than the place names and locations that dominate public discussion. Locations are the ABCs of geography, and they must be composed into the geographic equivalents of words, sentences, and paragraphs to reveal their meaning. The real language of geography lies in the relationships that tie places together, make each place unique, and render the world a more interesting place to live and explore.

It is disturbing when students can't locate countries on world maps, just as it is disturbing when they do not know the alphabet or multiplication tables. Far more troubling, however, is the fact that most Americans, even highly educated people, do not know the purpose and function of geography. The pollsters have focused narrowly on a "states-and-capitals" view of geography. Almost without exception, the tests have emphasized place name recognition and location. They ask, "Where is a place?" but almost never ask, "What is its climate, culture and economy?" They never ask, "Why is it there?" or "How is it connected physically, socially, and economically to its neighbors, its region, its world?" Some tests have erroneous answers. Many are regionally biased. In one poll, for example, 60 percent of the questions were about the Western U.S., and the pollsters concluded that Eastern students don't know geography as well as Western students. Only occasionally do the tests ask questions that require an understanding of geographic relationships among places, and they never require an interpretation of interactions of multiple phenomena in three-dimensional space.

To experience the geographer's perception of all this, imagine that you are watching the nightly news report on illiteracy in the United States. As the reporter speaks, it becomes apparent that they know only about the ABCs; with no knowledge about words, sentences, and paragraphs. The revelation would be frightening, and you might well despair of hope for widespread literacy in a society whose major spokespeople knew so little about language. From the geographer's perspective, this scene is repeated frequently as pollsters, journalists, and assorted other spokesmen decry geographic ignorance based on a definition of geography that greatly understates its worth.

For two millennia, geography was recognized and valued. In modern terms, we might say that it was a leading recipient of government research funds. Its body of knowledge was considered so valuable that some of the old and most of the new information was classified and held under the tightest of security by governments and mercantile companies.

The explosion of knowledge during the Renaissance greatly increased the complexity of geographic study and pointed toward specialization. Most of the familiar scientific disciplines of today arose during the following centuries, and many of them descended directly from roots in geography. After that time, geography as practiced by a small group of specialists grew more and more scientific, but it lost its preeminence in the broader realms of science and industry.

No doubt a prime reason for the decreased status of geography after the Renaissance was its inability to deal with the spatial complexity of real world phenomena in a rigorous fashion—a chronic problem that became critical with the proliferation of information brought on by the Renaissance. Geographers were forced to choose between excessively general studies of large areas, on the one hand, and detailed studies of excessively small areas, on the other.

In recent decades automation has enabled geographers to regain some of their pre-Renaissance holism.

Geographers responded to the impetus of new information technology when computers became available in the 1950s. But the "Quantitative Revolution," geared to the functions (primarily mathematical) that computers would perform well at that time, lacked the graphic, visual richness of traditional geography.

The improved graphics capabilities of the 1970s, were quickly grasped by geographers. Since then computer cartography, remote sensing, and quantitative spatial modeling have progressed along separate lines, and they are just beginning to merge in the 1980s and 1990s as automated geography.

In the midst of this digital revolution, automation has forced developers and practitioners to recognize explicitly what geographers have practiced intuitively for centuries. They are finding that geography, as a scientific and technical enterprise, is far more difficult than it looks.

A typical GIS employs the most challenging technologies of the Information Revolution: digital remote sensing, computer cartography, computer graphics, artificial intelligence, quantitative spatial modeling, spatial statistics, and spatial data structures—tools and methods that were barely understood by most practitioners as little as a decade ago. The data structures are, in fact, considerably more complex than those employed in computer-aided design (CAD) systems for engineering applications.

Geography certainly isn't what it used to be. For centuries geographers were explorers, and they excited the world with what they found. Today the thrill of discovery comes not from new places but from new knowledge about familiar places. And that knowledge is so varied that a confusing array of specialties results. Many geographers define their fields as the science of spatial relationships, with emphasis on the Earth. Beyond that commonality, they quickly part company.

Some focus on an individual phenomenon and examine its distribution through space. (As a nation, we are willing to pay large sums for government agencies to do similar studies: the Census Bureau with population; U.S. Geological Survey with land and water features; National Institutes of Health with the incidence of lung cancer.) Other geographers are more interested in the relationships that cause such features to vary through space and time. (How, for example, do the distributions of rainfall, soils, roads, and industry affect the distribution of population?) Others focus on a place and study the interrelationships that occur there. Still others examine the physical and cultural character of a region that exists because of the area's unique combination of geographic features.

With these skills geographers contribute powerfully to decision making, research, and analysis of environment, resources, transportation, emergency management, facility siting, logistics, military intelligence, and many other topics of vital interest at local, state, national, and global levels.

These changes in the function and practice of geography have brought a new dimension to the satisfaction and excitement it offers, which might best be explained with a personal illustration.

Once on a remote mountainside in upstate New York, I compared the landscape before me with satellite data in my hand, and suddenly realized that the forest was undergoing massive changes over hundreds of square miles—changes which, in theory, might be related to the processes that cause acidification in some Adirondack Mountain lakes.

For the next three years my colleagues and I conducted research, examining the Adirondack case in detail and discovering similar processes in Sweden and elsewhere. Finally we concluded that the suspected relationship was correct and that it could have profound implications regarding the impacts of acid deposition. The elation I felt at the initial moment of discovery could not have been greater if the finding had been an uncharted island, and the research was as intriguing as any mystery could have been.

It is, of course, more difficult to share this type of discovery with others than to share the discovery of foreign lands, but, when the tale is told, many listeners seem fascinated by it.

Geography, like mathematics, is a part of everything. It can be as common as navigating a neighborhood street or as complex as matching billions of bits of data in three-dimensional space. At the high end, we now depend on computers, satellite sensors, and GIS to collect and manage the massive databases required to describe large areas in fine detail. We build computer models to simulate the workings of real world systems that transcend the usual division of physical and social sciences. Often we integrate the work of more specialized disciplines and paint the "big picture." Often, through our science, we strengthen the linkages between other sciences and the real world.

Once again geographers are contributing heavily to the broader scientific community. The status of the discipline is improving, and the general public is gaining substantially in its appreciation for and understanding of geographic phenomena.

There is hope for better understanding through the promotion of geography. Congress has declared a National Geography Awareness Week, and the National Geographic Society is sponsoring a campaign to promote geographic education.

As a professional geographer I applaud these efforts. At the same time, I ask, "Which geography will we recognize and teach?" Yes, the states and capitals and thousands of other locations are important. Physical and cultural features are important. We must promote this fundamental aspect of geography, but we also must move beyond to a broader understanding of scientific and technical geography. If we can share that broader view, people will demand not only that geography be taught in our schools but that it be practiced more diligently in science and public policy as well.

Geographic Intelligence

Geographic intelligence is closely aligned with spatial intelligence. Geographic phenomena are described in terms of geometry, topology, and attributes. Geometry is concerned with spatial form; topology is concerned with spatial relationships; attributes are the quantitative measures and qualitative descriptions of earth phenomena that define the content of that space.

Gardner (1983) suggests that there are seven different kinds of intelligence: linguistic, logical-mathematical, spatial, musical, bodily-kinesthetic, interpersonal, and intrapersonal. In Western parlance "intelligent" people are those who excel in the first two. (Note, for example, that many of the qualifying tests for Mensa membership are based on combined "Verbal & Math" scores.) Yet, it would seem evident that all seven are important to some degree in determining the success of individuals within our society. All contribute to the collective progress of our society.

Spatial intelligence, in particular, is essential in activities as diverse as designing a complex machine, planning a transportation system, or programming a space mission. Any undertaking that involves location, movement, or integration is likely to require spatial or geographic intelligence.

Clearly, spatial intelligence is undervalued and poorly recognized in current western culture. We have come to accept this as a normal state of affairs, but it hasn't always been so. Prior to the Renaissance, geography and astronomy—both spatially oriented disciplines—were preeminent sciences (Boorstin, 1983). Why does current Western society value linguistic and logical-mathematical intelligence more than the other forms of intelligence?

Historically, innovations in information technology have had profound impacts on societal interests and capabilities in the areas of intelligence listed by Gardner. Gutenberg's printing press, for example, prompted a burgeoning of linguistic intelligence and the computer, per se, prompted major advances in logical-mathematical intelligence. There is reason to suspect that the impact of the new digital information technologies may affect geography as profoundly as did the printing press. Indeed, it may even be that these technologies are prompting a burgeoning of spatial intelligence similar to that of linguistic and logical-mathematical intelligence in the past.

The vision of a new scientific milieu in which spatial intelligence operates on a par with linguistics and mathematics is an exciting prospect. The challenge for the scientific and technical community of today is to achieve widespread geographic literacy in considerably less than the 500 years that it took to achieve widespread linguistic literacy.

(This column originally appeared in the June 1990 issue of the The Mensa Bulletin. *It is reprinted here with permission.)*

Geography is to GIS What Physics is to Engineering

By Jerry Dobson

GIS has been an easy sell because of its bright promise bolstered by impressive demonstration products, but many key managers and scientists are beginning to question the promise. Applications, scenarios, what-ifs, problem solving, spatial analysis, geographic analysis—these are pat answers to the dreaded question, "What are you going to do when your GIS is done?" When the system is working and the databases are prepared, what are you really going to do that you couldn't have done before? More important, how are you going to do it?

As I read Stan Openshaw's comments about the lack of analysis in GIS (*1990 GIS WORLD Sourcebook*), I was reminded of recent conversations with one friend in business and another in academia. The businessman is founder and president of a successful applications software company. He has a good appreciation for geography and decided to improve his mapping software by adding GIS features. He asked if I could steer him toward people who have actual experience using GIS for analysis. Everyone he contacted was still in the building stage and could not help him. The other friend is a professor at a university geography department that has a faculty opening in GIS. He complained that the applicants know how to make a map or do an inventory, but not how to analyze geographic information. His department elected not to hire any of the applicants that year.

The promise of GIS is brighter than ever, but the burden of proof is upon us. The time has come for a major step forward in GIS functionality, matched by a similar leap in user understanding and skills. Our objective should be an analytical GIS that will support the analysis of:

- The distribution of an individual phenomenon through space. This could be anything for which location is important—a mineral resource, a utility network or a population characteristic.
- The interrelationships among two or more phenomena that cause their distributions to vary through space and time. For example, do population and income distributions interact with transportation systems to determine retail market potentials?
- The interrelationships that occur among phenomena because of their coincident location at a specific place or places. Focusing on an Adirondack Mountain

watershed, for example, what unique combination of acid precipitation, landscape, land use, and hydrology causes lake acidification to occur?

• The physical and cultural character of a specific region or landscape that exists because of the area's unique combination of spatial phenomena. This includes regional geography, but it can include efforts as pragmatic as those of Chambers of Commerce trying to define and promote their districts for tourism or industrial development.

The ultimate GIS also would support development of the theory, methods, models, and techniques required to conduct analyses of the types described above.

To meet this analytical challenge, we in the GIS field must draw upon the tradition of geographical analysis that has gone before us. As bitter as it may be to some, we will all have to learn to say the G word—geography. No discipline has a monopoly on any type of analysis, but for centuries geography has been based on the same types of information and the same types of inquiry that the future of GIS will depend. Geography offers the GIS field the insights of 2,500 years of such inquiry, and the methods, techniques, and theories that have derived from this struggle. In return, GIS offers a digital model of the real world and a digital version of the tools geographers always have used to represent and study that world.

As both a geographer and a GIS specialist, I often have puzzled over the uneasy relationship between geography and GIS. They need each other, and their futures are permanently linked. Yet, for some unfathomable reason, many geographers look on GIS with condescension, and many GIS specialists have an almost pathological aversion to the word geography. Both camps may be surprised to learn that the functional description of GIS offered above has been published more than once as a functional description of geography. What seems to divide the two fields, more than anything else, is the age-old division of theory and practice.

Geography is to GIS what physics is to engineering. Engineers can build a lot of bridges without going back to the first principles of physics, but they better not violate them; bridges will fall and people will be hurt. Similarly, GIS analysts can digitize massive databases, integrate disparate data and conduct certain types of spatial analyses without referring to the basic principles of geography; however, the analysts who violate these basic tenets eventually will lose money, credibility and clients.

Geography and GIS are every bit as intertwined as physics and engineering. With advanced machines, like the super collider and advanced GIS, the frontiers of theory and practice go hand in hand. There is no room for condescension at that level. The same holds true for GIS applications, which can include research at the frontiers of geography and other disciplines. Even at its most basic level—digitizing databases, making measurements, producing maps and overlays—GIS is an intellectual enterprise requiring substantial knowledge of geography, cartography, topology, and many other fields.

When Thomas Wikle surveyed GIS users in private business, government, and academia (*1990 GIS WORLD Sourcebook*), not a single respondent listed geographic methods as a necessary college course. As GIS moves to its next analytical phase, surely geographic methods will move toward the top of the list of required courses.

For their part, academic geographers are moving to fill the gap. According to the Association of American Geographers (*1990-91 Guide to Departments of Geography in the United States and Canada*), 156 geography departments (65 percent of the total number in the two countries) claim GIS as a departmental specialty. [Update: Subsequently, GIS became a key specialty in every major university in the country, the GIS Specialty Group became the largest specialty group in the AAG and some 70 universities joined the University Consortium for Geographic Information Science.]

I suspect that GIS specialists will come to think of geography, not as the property of a specific academic discipline, but as one of many skills, like mathematics and computer science, that are essential to GIS. Conversely, it will be harder for geography to claim a unique position in the analysis of geographic and other spatial phenomena.

The "G" in GIS is there for a reason.

(This column originally appeared in the February 1991 issue of GIS World.*)*

Real Survivors Wrote Geography

By Jerry Dobson

For thousands of years, explorers have searched for unknown places, and their bitter plights sometimes show in the names they left behind: Hopeless, Desperation, Disappointment, Disenchantment and Dismal. For others, discovery came as a byproduct of seeking fortune, making a harsh living or stumbling off the beaten path. Those who survived brought back priceless knowledge that resounds to the everlasting benefit of society. Those who died kept their secrets, of course, unless delivered by companions or discovered by later seekers.

Suddenly in summer 2000, popular culture discovered survival, or rather a pathetic imitation of it, in a television series called "Survivor" (CBS). I didn't watch the show, because the personalities pre-selected to compete were such a turnoff, but I read enough to know that the principles espoused are precisely the opposite of what has worked for centuries among countless real-world survivors.

Geographers, in particular, are modern inheritors of the real-world tradition of exploration, hardship and survival. It's no coincidence, therefore, that "Survivor" is cast as a geographic challenge. The first trial was the classic (trite?) deserted island, and the next campaign will be in the arid "Outback" of Australia. The settings may be real, but the show's contrivances cater to the popular misconception that survival depends on rugged individualism, ruthless cunning and athletic prowess. Real-world experience, however, favors benevolent leadership, personal sacrifice and endurance born of sheer will.

Dog Eat Dog

Oddly, this misconception prevails at a time when accounts of real-world survival are popular and widely available to dispute it. The wreck of the Essex (Philbrick, 2000), the Galveston Hurricane of 1900 (Larson, 2000), Ernest Shackleton's expedition of 1914-1916 (Worsley, 1977) and the disastrous climbing season of 1996 on Mt. Everest (Boukreev, 1999; Krakauer, 1998) starkly contradict the TV show's dog-eat-dog philosophy.

Indeed, the social dynamic among the "Survivor" cast bears an uncanny resemblance to that of the Donner Party, the quintessential model of failure in the wild (Stone, 1956). Dissension, banishment and lack of leadership delayed their progress across the plains. When disaster struck in the high mountain pass that now bears their name, death rates were highest among the very old and young as well as strapping young men—who died most often simply because they didn't have anyone to take care of them. Most likely to survive were women and members of large, cohesive families who took care of one another. So much for rugged individualism and athletic prowess.

For selfless leadership, the one great truth that trumps all others lies in Shackleton's Antarctic expedition, universally recognized as the greatest endurance of all recorded history. On Jan. 18, 1915, Shackleton's ship, aptly named the Endurance, was beset in the solid ice pack of Antarctica. The only hope for him and his 28 men was to ride the ice on faith that it would eventually spit them back into the Atlantic Ocean. Nine months later, the ship broke up and was abandoned. For another five months, they lived directly on the ice. They sailed in small boats to frozen, uninhabited Elephant Island, where most of the men stayed. Shackleton sailed with a small crew to South Georgia Island and then led four separate rescue attempts during a four-month period. In the end, not a man was lost, and Shackleton was recognized as the greatest expedition leader of all time. Accounts written by his crew cite Shackleton's fairness as exemplary and his care for each man as almost motherly. So much for ruthless cunning.

And what about the TV show's practice of banishing fellow survivors? In real-world ordeals, that happens with a reluctance which borders on pain and often ends in tragedy. During John Wesley Powell's first exploration of the Grand Canyon, three men gave up and struck out overland on their own (Wilford, 2000). Powell supported them with food and guns and described their parting meal as "solemn as a funeral." One day later, he and five remaining men emerged from the canyon, but those who left died on the plateau above.

A similar, perhaps greater depth of anguish was expressed by the men of the Essex, the ship that was sunk by a whale, inspiring Herman Melville to write *Moby Dick*. In an arduous journey back to South America, three fellow sailors elected to remain on Henderson Island. The crew's tragic fate can be attributed to errors of judgment caused by two factors: 1) geographic ignorance of islands that lay to the west and 2) the captain's failure to act on his own convictions. According to author Nathaniel Philbrick, Democratic consensus doesn't work well in survival situations; it's better to select the most competent leader and stick with his or her judgment.

A Price to Pay

In today's comfortable world, we sometimes forget the price we used to pay for exploring and subduing new lands. Incredibly, after the Challenger space shuttle accident, a few scattered voices actually called for an end to the U.S. manned space program. But anthropologist Ben Finney, a noted authority on Polynesian navigation, estimates that as many as 250,000 Polynesian sailors may have lost their lives at sea, searching to find the islands that survivors eventually settled (Finney, 1994).

For sheer intrigue and personal risk, no geographic episode surpasses the exploration north of the Himalayas ordered by British commanders and carried out by Indian agents from 1865 through 1882 (Wilford). When the "Survey of India" was complete, surveyors looked farther north toward lands controlled by Russia and China that were closed to outsiders. Observing, however, that Indian natives crossed freely into the forbidden lands, they eventually recruited Indian educators Mani and Nain Singh to make the journey. The two suffered severe hardships, but returned with invaluable information, and Nain Singh eventually received a gold medal from the Royal Geographical Society. Many agents were less fortunate—some murdered, some robbed and at least one sold into slavery—all for the sake of advancing geographic knowledge and thereby tipping the balance of power among Great Britain and its enemies.

Collective Survival

Few people ever will face the kind of individual ordeal that "Survivor" is meant to simulate, but disaster hasn't gone out of style. Natural hazards, technological accidents and wars still threaten communities around the world. For lessons in collective survival, read Eric Larson's account of the Galveston Hurricane of 1900. You'll find plenty of admonitions against over-reliance on scientific theory, blind trust in technology and bureaucratic abuse of power. You may empathize with Isaac Cline, the meteorologist on site, only to realize that he was one of many weak links in the flawed chain of information that failed to save 8,000 lives, including that of his own wife. Along the way, you may appreciate Larson's account of the painful birth of the National Weather Service. He points out, for instance, that the service couldn't have existed before the telegraph was in place to collect simultaneous readings and distribute predictions in advance of predicted events. Information technology was as important then as now for protecting communities and preventing collective tragedy.

Did "Survivor" send the wrong message? That question has been asked by numerous commentators following the show's startling success. The answer is "no." It's the wrong message, but television didn't send it. We, the people, formed a misconception based more on bureaucratic and political shenanigans than on life-threatening ordeals. We sent that message to ourselves and affirmed it in thousands of water-cooler conversations after each broadcast. That's an odd direction for an audience "surviving" in the most affluent society in the history of the world. For Americans, at least, survival is easier than it has ever been.

Ironically, bad behavior, such as that portrayed on "Survivor," is a luxury we can afford only because the means of survival presently are so plentiful. If and when we face real tests, we'll have to fall back on good old-fashioned values of benevolent leadership and personal sacrifice or suffer the consequences.

(This column originally appeared in the November 2000 issue of GeoWorld.*)*

Did Albert Einstein Really Say Geography is Harder Than Physics?

By Jerry Dobson

"As a young man, my fondest dream was to become a geographer. However, while working in the Customs Office, I thought deeply about the matter and concluded that it was far too difficult a subject. With some reluctance, I then turned to physics as an alternative."
– Albert Einstein (unpublished letters)

Part I – Question

No doubt Albert Einstein (1879-1955) was smart enough to figure this out, but did he actually say it?

Jeff Osleeb, chair of the geography department at Hunter College, City University of New York, mentioned that he was aware of such a quote and wondered if it could be found and verified.

Curiosity prompted me to seek an answer on the Web. Using the Alta Vista search engine and the search command "Einstein AND geography," I quickly found three instances of the full quote and later a fourth. Two of them were on geography department Web sites; two belonged to individuals. One pair contained precisely the same words as those above, and the other two differed only slightly, ending with "a substitute."

I was gratified to find the statement, but troubled by the vague reference to "unpublished letters." When citing a letter one ordinarily would state a recipient and a date. Also, it seems odd to say "letters" in plural. Is it from a published collection, or did Einstein repeat the same statement to several recipients?

Searching for Clues

To investigate the matter further, I sent an e-mail inquiry to each of the Webmasters. At the first geography department I checked, at a Canadian university, a former student had prepared the Web site. At first, no one recalled the quote's origin, but the Webmaster took an interest and made inquiries among his colleagues. The source was a faculty member recruited from another Canadian university. It had been thrust on him by a fellow geographer specifically because of family connections with physics. I contacted the professor who had done the original teasing. He got it in 1994 or 1995 from the newsletter of an American university just across the border. I contacted the newsletter editor. His source was a professor in his own university who found the quote on a sign in the physical geography laboratory of another American university. That trail led to a former department chair, a respected colleague and good friend of mine, who didn't recall the original source.

At the second geography department I checked, neither the Webmaster nor the department chair could recall the source. I won't mention names because I don't want to embarrass anyone (unless, of course, it happens to be a football rival who beat the University of Tennessee, dashing all hope of a national championship in 1997). The first individual said he got it from the bulletin of a state geographical society in 1994. I called the bulletin's editor, and she said it was sent in by a teacher whose name had since been forgotten. The second individual said she retrieved the quote from a fellow student who doesn't remember where he got it.

Simultaneously, I searched the Web to see whether I could find the letter itself. As far as I can determine, there's no electronic source for the collected writings of this important man. The Albert Einstein Archive is at Hebrew University, Jerusalem, and a photocopy collection is held by the Princeton University Library, Princeton, N.J. If you're in either neighborhood, please check for me. That was as far as I could go at the time without turning the exercise into a major quest.

Internal Consistency

Still unsatisfied, I asked myself: Is there anything in the quote itself that could serve as a test of authenticity? What about Einstein's employment? Did he really work in a Customs Office, a place likely to prompt thoughts of faraway places and, perhaps, geography? No, he worked in the Swiss Patent Office, a place better suited to the contemplation of physics. Furthermore, by that time he had completed four years of study in physics and mathematics at the Federal Polytechnic Academy in Zürich. By 1905 he had published five major papers on physics while working at the Patent Office. My impression is that his field was firmly chosen before he took the job.

Earlier, however, Einstein attended the Luitpold Gymnasium in Munich, where at age 12 he devoted himself to "solving the riddle of the 'huge world'"—a possible reference to geography. And, yes, he must have found the subject difficult because his grades were poor in history, geography, and languages, and three years later he left school without a diploma. If young Albert really considered and rejected a career in geography, it must have been during this period rather than during his years in government service.

Physics Envy

Let's be honest. We geographers are a bit credulous about this quote because of what one writer has called "physics envy." Physics has enjoyed a status and recognition that our discipline hasn't experienced since the early Renaissance.

I work in a highly technological place where geography is appreciated, but physics, chemistry, and engineering dominate. Our offices, for example, have metal walls originally designed so "hard" scientists could blow themselves up with minimal collateral damage. Yet early in my career I once took part in a minor triumph of geography over physics. We completed an extensive siting analysis and produced about 40 maps illustrating the options. With scores of magnets, we stuck those maps on my wall to get an overview. A physicist friend happened to walk in just as someone joked that we ran the risk drawing all of the neighboring offices' metal furniture to the wall. "Oh, no," the physicist said without a trace of amusement, "the magnetic flux just goes into the metal and back to the magnet. Can't possibly pass through!"

Just then it occurred to me that my neighbor was moving, and I decided to take a chance. I led the crew into the hall and opened my neighbor's door. Sure enough, all his furniture was crowded against the back side of my wall. "What can I say?" the physicist said.

Geographers take pleasure in small victories.

Part II – The Answer to an Ongoing Investigation

My initial column elicited numerous communications from readers, many with excellent leads and suggestions and now, the 20-year-old mystery has been solved. The clincher, was a surprising E-mail message from the person who fabricated the "quote" out of whole cloth. It was none other than Duane Marble, a renowned pioneer of the GIS revolution. He wrote to tell me the origin and circumstances behind a practical joke that took on a life of its own. Here, paraphrasing Marble, is the true story.

The Truth, the Whole Truth, and Nothing But the Truth

Marble is now a Professor of geography at The Ohio State University, Columbus, but the episode began during his days as a graduate student at the University of Washington, Seattle, and as a young faculty member at the University of Pennsylvania, Philadelphia; Northwestern University, Chicago; and State University of New York at Buffalo. In those years, he worked closely with Michael Dacey, who had read a lot of physics literature and concluded that the problems addressed by physicists, geographers, and astronomers were differentiated more by scale than by difficulty.

Several years after Marble arrived at Buffalo, the state of New York built the Amherst campus. The physics department was favored with a lavish, four-story building, whereas the geography department languished in rented, temporary space. When the geography department finally moved on campus in about 1977, it was to the top stories of the "Physics Building." Marble's office and the GIS Lab were on the third floor, where geographers daily confronted the displaced and disgruntled physicists. Recalling Dacey's comments, Marble composed the "Einstein" statement and posted it in plain view of the former tenants. He says, "It was very popular with a segment of the geographers, and it seems to have spread far and wide since that time."

Marble was just having fun, not trying to fool anyone. His reference to the "*Customs* Office" rather than the Patent Office was a deliberate clue, as was the citation: His remaining copy says "*very* unpublished letters."

Hard Sciences, Difficult Sciences

At first, I was surprised to learn that the real author was a GIS pioneer. In retrospect, I've decided it's logical. For 2,500 years, geographic analysis depended on the visual and intuitive skills of rare individuals who could "see" spatial patterns and discern their meaning. Only in this century, when Michael Dacey, Duane Marble, Waldo Tobler, Roger Tomlinson, and other trailblazers tried to mathematize and computerize those spatial skills, did we truly appreciated how difficult such skills are.

Consider the quote, stripped of its attribution, and ask yourself: Is there any truth to the underlying premise that geography is more difficult than physics? In answer, I once wrote, "For most of this century, society has been preoccupied with the 'hard' sciences; GIS may help us move beyond to tackle the difficult sciences, as well" (Dobson 1993). Indeed, I believe "hard" answers are possible in physics, chemistry and similar sciences precisely because *relatively* simple processes can be identified and isolated from the complex interactions of real-world landscapes, thereby reducing uncertainty. Conversely, social sciences are "soft" because their subject matter is too difficult for such "hard" answers. Geography is even more difficult because of its need to integrate physical and social processes in complex, 3-D environments operating over time.

Consider the following example (Dobson 1991): If a physicist marks a single point on Earth's surface and conducts some type of physical process analysis, he or she is doing physics. If the physicist does the same thing for a thousand points, it's still physics. But, if the physicist maps (or enters into a GIS) the distribution of the answers along with a lot of other information and tries to understand how the physical process is influenced by climate, geology, human activity, and other relevant factors, then he or she will be doing geography. The physics will remain, but the geography will be as challenging as the physics.

I have tremendous admiration for physicists, and my remarks shouldn't be taken as disrespect. The physicist mentioned in my January column was Jeff Jalbert, my good friend and colleague then on leave from Denison University, Granville, Ohio. We co-authored a report on cell-based screening using GIS (Jalbert and Dobson 1977). We share a mutual respect that I wish were more common among the disciplines as a whole.

A Hidden Theme

A hidden theme in my January column was verity on the Internet. We must establish, in its youth, that the Internet is a powerful broadcast medium demanding the same verification and documentation expected in print media. Academic institutions, especially, should set high standards and lead by example.

To be fair, the bogus quote had appeared in two print publications, one a refereed journal, before appearing on the Internet. That's disappointing. Maybe the temptation is just too great when a quote, no matter how spurious, seems to place the greatest physicist of all time in our court.

Part III – Epilogue

I never set out to be a debunker, but 13 years of writing for *GeoWorld* conditioned me to check every fact in everything I say or write, no matter how casual it may seem at the time. The habit has served me well, I think, and here is a recent example.

There is an observatory on the roof of Lindley Hall, home of the Geography Department at the University of Kansas, and there is a popular campus myth that Pluto was discovered from it. I heard the story shortly after I arrived on campus, checked the Internet, and quickly found that it was not true. Clyde Tombaugh actually discovered Pluto while he worked at the Lowell Observatory in Flagstaff, Ariz., in 1930. Clyde, a native Kansan, was only 24 years old at the time. He then came here to continue his education, and the observatory bore his name for many years. So, I thought, it may, at least, be true to say that the discoverer of Pluto used that observatory. Lately, hearing news stories about Clyde's ashes departing for Pluto and reading Dava Sobel's new book, *The Planets* led me to dig deeper. Alas, Lindley Hall wasn't built until 1943, and Clyde completed his degree in 1938. The only remaining vestige that Clyde would have touched is a six-inch refractor telescope still mounted but unused in the abandoned dome visible from my office. I sent a debunking e-mail to my fellow faculty members, and it was gratifying when Kevin Price recalled the Einstein episode.

Other geographers mention it to me from time to time, but the Internet as a whole has not been so kind. Far too many Web sites still use the quote without qualm. Initially, I was gentle with those who did not heed my "hidden theme" about verity on the Internet. Now, however, any fact can be tested easily and quickly on the Web and the truth has been known for nearly a decade. So, I say shame on them, especially the university departments and individual faculty and graduate students.

Today, about 150 Web sites mention the quote, and about half of those acknowledge Duane Marble as the real author. Less than a dozen acknowledge me as the sleuth who debunked it.

Was Einstein Ever a Geographer?

Well, it turns out Einstein (1954) did write one paper on the causes of stream meanders.

Ah, but did he get it right? That should be the true test of whether geography really is tougher than physics or not.

(This column combines two columns that originally appeared in the January and May 1997 issues of GeoWorld.*)*

Musings on the Revenge (and Forgiveness) of Geography

By Daniel Sui

In 1997, *The Economist* pronounced distance dead based on the astonishing development of telecommunications led by the Internet, cellular phones, wireless sensor webs, etc. If I can paraphrase Mark Twain, the "death of distance" announcement proved to be greatly exaggerated.

In 2003, six years after the announcement, *The Economist* published an article on "the revenge of geography," in which the news magazine subtly conceded that its predictions on the role of distance and geography in the information age were misguided.

Distance Resurrected

As a geographer, I applauded the article's publication and forwarded it to my non-geographer friends. My overly enthusiastic reaction was primarily due to the fact that geography rarely receives extensive coverage by well-established mainstream media.

Furthermore, the article "drove home" the message that it was naive to think that the Internet's global reach would make geography irrelevant. On the contrary, wired and wireless technologies have bound the virtual and physical worlds closer than ever.

According to the article, distance was resurrected, and locational information has become a commodity. The revenge of geography was so profitable that a new billion-dollar industry, location-based services (LBSs), emerged in the first few years of the 21st century.

With computers increasingly embedded in our surrounding environment (and sometimes in our own bodies), LBSs are quickly becoming an integral part of ubiquitous computing (ubicomp) and ambient intelligence with context-aware capabilities. The demand for LBSs will continue to expand, and the market has already penetrated a variety of issues that range from the global war on terrorism to intelligent transportation systems to monitoring individual physical activities.

The technological advances in acquiring, transmitting, analyzing and communicating locational information made it feasible to geocode and track almost everything and everyone on Earth. For example, cell-phone companies now offer services that track the physical location of other cell phones.

The development of sensor webs, along with devices such as "smart dust," enabled scientists to acquire environmental information in real time. Although it will take some time to sort out the long-term implications of these developments, LBS and ubicomp will fundamentally change how we manage and process spatial data.

More than Location

With the maturity of ubicomp, sensor webs and the embedding of various location-acquisition technologies in the environment, we'll be further flooded by terabytes and exabytes of data with greater spatial and temporal resolution. The late Peter Gould's vision of a "spatial century" in the 21st century has technically materialized.

Our sense and sensibility about the world will hinge, fortunately or unfortunately, on how well we can absorb the huge volumes of data. This provides ample opportunities and daunting new challenges for geography as a discipline.

Having recently re-read Marshall McLuhan's earlier studies on media published in the 1960s, I realized that media can promote and kill the subject it covers. Despite the global publicity it gave to geography, the "revenge of geography" article and related media coverage on LBSs also generated some serious concerns and anxieties due to the tendency to subtly equate geography to location.

Previous geography literacy polls, which received a lot of media coverage in the late 1980s and early 1990s, raised the public awareness on geography but also unintentionally strengthened the public perception of geography as a trivial discipline of place names. Now there's also a danger of equating location (or LBS or GIS) to geography.

Breadth and Depth

Geography is more than place names and locations. I take pride in being a geographer, primarily because geography is a diverse and intellectually vibrant discipline—this becomes clear as almost all approaches to scholarship (science, engineering, social science, humanities, etc.) are accepted and practiced by geographers.

To experience the breadth and depth of the discipline, I suggest (especially to those in the GIS industry who aren't geographers by training) attending an annual meeting of the Association of American Geographers and "hanging out" with geographers for a few days. Or look at the table of contents of the *Geography in America at the Dawn of the 21st Century* (Gaile and Willmott, 2003), which should give non-geographers a good overview of the diverse threads of geography's intellectual tapestry.

Throughout geography's history, no technological innovations have affected its development like GIS. Undoubtedly, GIS has transformed geography as a discipline in many fundamental and positive ways, but the interaction between GIS and geography is a two-way street. In recent years, many conventional geographies have been revitalized in the context of GIS and GIScience.

For example, geographers recently discussed Waldo Tobler's first law of geography and its relevance to GIScience, Torsten Hägerstrand's time geography and the development of people-based GIS, new locational theory in the context of computing, and the implications of Yi-Fu Tuan's humanistic geography for studying people's sense of space and place in the information age.

Moving beyond the technical/methodological realm, geography and GIS are more tightly intertwined intellectually, and this trend will continue as various technologies are converging toward location or context awareness. If geography is to GIS what physics is to engineering (as my colleague Jerry Dobson articulated in 1991), I'd like to challenge everyone to think about the short- and long-term implications for GIS applications if geographic laws and principles are ignored or violated?

Moving Beyond Vengeance

In comparison to other "species" in the academic jungle, one of the virtues that geography can be proud of is its ample tolerance of a variety of approaches to studying Earth as home for humankind. And as an intellectual enterprise, geography is quite forgiving in spirit.

On close scrutiny of *The Economist* article, I think it's technology, not geography, that has taken revenge in recent years. To move forward, "vengeance" won't get us far; proceeding with a spirit of forgiveness is a better alternative, because, to paraphrase Hannah Ardent, "forgiveness is the key to true human freedom and action, as one can only forgive to the degree one loves."

Edward Tenner (1996) reminded us that all technologies are vengeful and tend to "bite back" in due time. The good news is that geography can potentially promote a spirit of forgiveness (if we can do it right).

It's this spirit alone—perhaps more than any technology—that may be the greatest contribution geography can make to humanity. Indeed, just as Josh Billings argued, "there is no revenge as complete as forgiveness."

(This column originally appeared in the June 2005 issue of GeoWorld.*)*

Geography's Public Image Impacts GIS

By Jerry Dobson

Despite the current boom in GIS and geographic information science, the Association of American Geographers and the American Geographical Society perceive an urgent need to improve geography's public image. Unfortunately, success in that endeavor is uncertain.

Geography's image obviously matters to these particular professional societies, but why should it matter to anyone else? The entire GIS industry would benefit by having a highly regarded discipline bear its flag. Consider, for example, that federal research and development (R&D) budgets presently are determined with little input from geographers or geographic information scientists and still less from the broader GIS industry. More important, GIS development *and* society's use of GIS need the theoretical and methodological underpinnings that geography has been developing for 2,500 years.

A Discipline Apart

Science and society rapidly are evolving toward the perspectives, methods and technologies that have been associated with geography and geographers for years. The greatest challenge ahead is to position the discipline so geographers may lead society into a new era of geographic capability and awareness. Success will depend on heightened recognition in the public arena, proactive leadership of science by geographers and enhancement of geographic institutions. Conversely, geography's poor image among the public, its peers, and politicians and administrators in charge of science policy diminishes present prospects. The present situation inhibits geographers as well as everyone associated with geography, geographic information science and GIS.

The "states and capitals" image of geography, so often formed in grade school, may sound comical, but it's pervasive, and its impacts are real. Geography, for instance, isn't listed as a separate, recognizable category of membership in the National Academy of Sciences (NAS). In shaping a national R&D agenda, chemists (197 members) and physicists (183), who have their own NAS sections, certainly have more influence than geographers (11) or geographic information scientists.

The American Association for the Advancement of Science lists geography, but it's oddly combined with geology (as on many college campuses), thereby illustrating another pervasive problem: The geographers' way of thinking is so different from the usual organization of science that no one seems to know where it fits.

Is geography a physical or social science? Even at this coarsest division, geography doesn't fit in either camp. This confusion derives from the fact that science, by and large, is fragmented into content-oriented themes. With its spatial orientation, geography cuts across the grain. That's integration, not fragmentation, but the diversity of topics appearing in publications and conferences misleads many topical specialists to label geography as fragmented. For all the lip service currently paid to integration, the scientific community doesn't tolerate integration very well.

Positive Steps

What can geographers do to ensure the discipline's rightful place in the broader scientific community?

- Identify wholeheartedly with GIS and geographic information science. GIS is a major driver in the societal movement toward geographic thinking. Geographers can't claim all the credit, especially for the technical accomplishments of computer scientists and engineers, but surely they should be recognized on a par with other contributors to the science and technology that bear their name.
- Start teaching geographic methodology much earlier in the college curriculum and even in the high school curriculum. Ask any geographer what defines the discipline, and he or she will insist it's not content but rather methods and perspectives. Yet undergraduate geography courses consist almost entirely of thematic content, and methods and techniques are reserved for graduate school. That's a bizarre contradiction that should be remedied as soon as possible.
- Campaign to ensure that geography is recognized as a scholarly discipline at the highest realms of science as well as on university campuses everywhere.
- Establish a commission to design and promote extensions to the scientific method to better serve geography and other complex earth sciences. Classic hypothesis testing works fine for individual processes that can be isolated and experimentally manipulated. It doesn't work well for complex earth systems. What's needed is a more encompassing and demanding method that tests for process, as before, but also insists on rigorous tests of spatial and temporal associations grounded in spatial arrays of real-world entities.
- Engage in the major scientific debates of our time, advancing new theories based on geographic inquiry and research. Only by engaging and occasionally winning such debates can geography establish its credibility and prove its value to the broader scientific community as well as the public.
- Enhance an existing journal or establish a new one to serve as the geographical counterpart of *Science* magazine. Publication in *Science* now serves as a hallmark of success in process-oriented research, but its editors and reviewers seldom understand or appreciate what constitutes good geographic research. The same can be said of *Nature* and certainly of specialized journals in other fields. As geographic capability and awareness spread, the scientific community and the public may come to appreciate an equivalent hallmark for holistic, spatial research.

None of these steps will be easy. It's a long-term building process starting with gradual acceptance and understanding of what GIS and geographers really do, geographers occasionally drawing blood in the great debates of science, reestablishing geography in Ivy League schools and someday winning ultimate recognition. Still, "the fight's the thing." If we passively accept our current status, we won't get to the first step much less the ultimate one. An activist discipline that falls short of its goals will be far better off than a passive one that rejoices in small victories.

The Popular Line

How odd it is that geographers can't excite the public about geography, yet others do so routinely. For example, Thomas Pynchon's *Mason & Dixon* novel and CBS' "The District" television show embrace geographic concepts that sound dry, technical and tedious on our professional tongues. Yet their writers somehow craft the surveying of a line, reckoning of longitude and GIS crime solving into alluring visions of tremendous appeal to huge, diverse audiences. Similarly, physiologist Jared Diamond constructed a new global geography in *Guns, Germs, and Steel* that is outselling, I suppose, the combined top-20 global geographies written by geographers. John Nobel Wilford's *The Mapmakers* makes cartography as intriguing as espionage. Charles Frazier's *Cold Mountain* is a masterpiece of landscape description.

Recently at Constitution Hall in Washington, D.C., my son Craig and I attended a Mark Knopfler concert. Knopfler's tour, latest CD and lead song are named "Sailing to Philadelphia." In the opening lines, Jeremiah Dixon proclaims himself a "Geordie boy" who measured up all Durham and Northumberland by his own hand. It was his fate from birth to make his mark upon the Earth. "Good surveyor" Dixon and "stargazer" Charlie Mason, who was born to chart the evening sky, are sailing to Philadelphia to draw the Mason-Dixon Line.

What enticed Knopfler to write and sing about two state lines laid out by surveyors more than two centuries ago and recognized henceforth by politicians and geographers? He grew up not far from England's "Geordie" country and may feel some pride in Dixon's American accomplishment, but something deeper must motivate his international audience.

What enticed Pynchon to write an entire novel on the same topic, and what captivates his readers? As a geographer, I'm intrigued by the technology used to produce the Mason-Dixon Line and by its later role in dividing North and South, but none of that comes through in the song or the novel. Pynchon imbues the line with mystical significance, but otherwise his *Mason & Dixon* mainly presents two lovable rogues doggedly discharging a surveyor's contract.

Listening to Knopfler's song and reading Pynchon's book, I laughed at the thought of what would have happened if professional geographers or surveyors had tried to excite the public about a survey line or, for that matter, any geographic feature on Earth or any method, technique or technology used to study them.

Why is it so hard for geographers themselves to stir public passions on the matters they hold most dear? It's high time to learn.

(This column originally appeared in the July 2001 issue of GeoWorld.*)*

GIS Has an Image Problem

By Jerry Dobson

To most insiders, GIS is an exciting world of unlimited potential to represent and investigate whole landscapes in ways that have never been possible before. Yet, to many others, GIS is a just an expensive way to make a map.

Better Than Virtual Sex?

In March 1996 a conference was organized specifically for the Department of Energy's national laboratories to inform another federal agency about emerging information technologies. I was there to promote GIS. The luncheon speaker on opening day was a gentleman who had spent 17 months on a National Academy of Sciences committee dealing with information technology, and the title of his address was "The Top Ten Information Technologies."

Beforehand, it occurred to me that, if GIS wasn't listed, I probably wouldn't respect his other 10 items much. Finally, he showed his top 10, and GIS wasn't among them. His next slide was entitled, "Some Things That Are Not So Hot," and there was GIS. I was stunned, as were many others in the crowd. When he finished speaking, I rose quickly to defend GIS, and I ended with a forceful statement: "If you pronounce the demise of GIS, you must say what will replace it, and nothing else on your list can replace GIS."

Later in the discussion, he turned to virtual reality and other ways of presenting information, beyond visualization, such as sound and touch. He described pressure gloves that will allow an analyst to feel the data in his hand. At that, someone from the host agency, said, "We get so much data thrown at us, you'll have to give us whole body suits."

To which the speaker replied, "Funny you should mention it. Our committee really did have to consider body suits, because one of the items on our agenda was virtual sex."

I stood back up and said, "Now, you've finally mentioned something that might replace GIS."

What's in a Name?

Yes, GIS has an image problem. I suspect the bad press occurs primarily because so many people have fossilized the definition of GIS at the level of functionality that was common about five years ago when GIS first hit the big time. That certainly seemed to be the case when the luncheon speaker tried to defend his assertion that GIS was "Not So Hot." Many people still think of GIS as a sort of spatial spreadsheet that merely represents 2-D surfaces stacked on top of one another like layers of paper maps. Some have dismissed the whole field without considering exciting developments recently released or under way.

A primary reason for establishing the University Consortium for Geographic Information Science (UCGIS) was a concern that the use of GIS as a tool is exploding faster than the understanding of GIS as a science. Indeed, one doesn't have to look far to illustrate this point. The Web home page of a major national firm, for example, touts its GIS services, but when the term is spelled out, it's a "graphical information system." Simultaneously, the flier for an upcoming GIS conference spells out "geospatial information system." I suppose someone might defend either term (There's no accounting for taste!), but the catch here is that both the home page and the flier were produced by the same "GIS" firm.

Counterpoints

My wife's grandmother used to have a saying, "Let's kill him and tell God he died." That's the first thing that went through my mind when the luncheon speaker put GIS on the "Not So Hot" list. But later I was able to muster a more rational (and legal) response when I spoke with him privately. I first showed him a graph of the remarkable growth rate of GIS documented in a GIS World Inc. Software Survey. Specifically, I showed him the number of new GIS products offered each year since 1975. The growth has been so rapid that the chart easily can be misconstrued as cumulative growth even when only the absolute number of new products is shown for each year.

Just in case he wasn't impressed by popular usage, I then told him about the National Science Foundation's early commitment to the National Center for Geographic Information and Analysis and its recent commitment to the National Center for Environmental Decision-making Research, of which GIS is a major component. I told him about 27 universities, one professional association, and one national laboratory forming the UCGIS specifically to advance geographic information science. Finally, I pulled my ace and said, "How many of your top ten have a full page in the *Whole Earth Catalogue*?"

Then I showed him the GIS page in the most recent edition. That did it! He genuinely repented and promised to remove the offending slide from future presentations.

I still hope to speak with someone in the "graphical-geospatial information system" firm. What I may say is that "graphical" just means "drawing," and I may ask what they plan to draw. Is it a map or is it something more? And if it's more than a map, wouldn't geographical be a better term? As for geospatial, it's not bad as a descriptive term, but it ignores or denies 2,000 years of intellectual development that have led us to this wonderful age.

A Fledgling Technology

In spite of its commercial success, GIS is still a fledgling technology. The functionality of today undoubtedly will seem primitive in as little as five or 10 years. We'll see better integration, especially with remote sensing, statistics, process models, and transport models. We'll expand to other dimensions, 3-D and time, not just for visualization but for analysis as well. We'll find better ways to represent culture and to deal with uncertainty. Spatial analysis will improve, but eventually will be only a small part of a unified place model for representing and analyzing complex landscapes. If the term "GIS" is allowed to grow with the field, the GIS of tomorrow will capture all of these processes and dimensions. Otherwise, we'll have to invent a new term, and we, ourselves, may forget that it was geography (I mean the science, not the academic discipline) that brought us here.

(This column originally appeared in the July 1996 issue of GIS World.*)*

What Can GIS Do to Save Geography?

By Nigel Waters

During March 14-19, 2004, the Association of American Geographers (AAG) will return to Philadelphia, the site of the organization's first annual meeting 100 years ago (*www.aag.org/annualmeetings/index.cfm*). Judging by the constant stream of "AAGeograms" that have been landing in my e-mail folder, the centenary will be a huge celebration of a century of North American academic geography.

It seems reasonable to ask, as geographers get ready for the occasion, just how well the discipline is doing in North American and other universities.

Glory Days

Academic geography flourished in the 19th century in the United Kingdom. I noted in a previous "Edge Nodes" column (Waters, 2003) how Alexander Maconochie, after returning from his governorship of the Norfolk Island penal colony in Australia in 1843, was appointed professor of geography at University College in London.

Other British university appointments soon followed—it was the era of the Empire and Darwin, geopolitics and trade, and survival of the fittest. Somehow geography seemed especially relevant.

On the other side of the Atlantic, Cambridge University's offspring, Harvard, had long provided geography with a warm and well-respected home. Geoffrey Martin (1998) described how the settlers of New England brought with them the geographical texts and concepts of European writers from Ptolemy to Varenius. These led to the development of geographical studies at Harvard and Yale almost from their beginnings. Established in 1639, Harvard included geography in its "List of Studies" by 1642.

In 1767, curriculum reorganization led to specialization by one of the four tutors in the areas of natural philosophy, mathematics, geography and astronomy. In 1803, admissions standards were raised, and proficiency was required in Greek, Latin, arithmetic and geography. As Martin notes, by the 1770s, sophomores were required to read William Guthrie's *New Geographical, Historical and Commercial Grammar*. In 1789, this was replaced by Jedidiah Morse's *The American Geography.*

Martin (1998) also noted that other New England colleges had early geographic offerings, including Brown (1899), Radcliffe (1904), Smith (1909) and Middlebury (1913). It's ironic that in 1892, Charles Eliot, president of Harvard, was placed in charge of the "Committee of Ten," of which a subcommittee was charged with developing the curriculum for school geography. A member of that subcommittee was William Morris Davis, also of Harvard, who was instrumental in founding AAG in 1904.

The Dark Years

American geography's dark years began in 1948, when the geography program at Harvard was summarily and arbitrarily eliminated by the university's president, chemist James B. Conant. It was, as the famed urban geographer Jean Gottmann stated, "a terrible blow to American geography" and one from which "it has never completely recovered." Neil Smith (1987) provides the definitive account of the bitter personal intrigue that led to the demise of geography at Harvard.

The decision at Harvard had far-reaching consequences that went further than removing geography for reasons of financial exigency. Conant stated unequivocally that the subject had no place within a university curriculum. These comments reverberated down the decades and have probably led to a string of other closures and academic departmental mergers.

Geographer though he is, Smith lays much of the blame with the discipline of geography itself, which he says was vulnerable then and that things haven't changed much in the ensuing 40 years.

At Harvard, geography claimed that its unique contribution was to provide a synthesis offering a unified vision of "man and the environment." The Harvard committee reviewing geography's legitimacy as a discipline was unmoved, because geography then, as now, was hopelessly divided between its physical and human branches. How then could it offer a holistic view of the world when its practitioners were so bitterly divided and contemptuous of each other?

Yale introduced a full-fledged Department of Geography in 1949, but it was eliminated within two decades. Strong, flourishing geography departments, such as at the University of Chicago, and small, outstanding departments, such as at the University of Michigan, have all gone the way of the Dodo bird.

Canada has fared no better. Departments of geography at the universities of Windsor, Concordia, McMaster and Alberta have all been merged with other departments. Academic geography also appears to be having challenges in Australia, where it has to come to various accommodations with environmental sciences.

It's surprising to me that few of those concerned have mourned the passing of their academic homes, and the Internet is strangely silent as to their former existence. They are gone and forgotten.

A White Knight?

Today, as in the mid-20th century, the academic discipline of geography appears to be under threat. This is true at locales as disparate as the United Kingdom and India. The state of the discipline in India is discussed in a review in Anu Kapur's book, *Voice of Concern* (see *www.tribuneindia.com/2002/20020825/spectrum/book7.htm*). In the United Kingdom, Nigel Thrift (2002) has taken an optimistic view of the discipline's future. Without ever mentioning GIS, he argues that the discipline has a strong future, but he sees it as a future in which human and physical geography go their separate ways.

In response to Thrift's paper, Ron Johnston (2002) argues that geography will not survive such a schism. Johnston appears to have learned the harsh lesson taught by Conant 50 years earlier and elucidated by Smith. Johnston says that geography will survive because 1) it attracts a lot of students, 2) it's intellectually vital and alive, and 3) it provides students with a marketable skill: GIS.

Johnston feels that this is enough, but perhaps it's not. An additional, more recent debate has taken place in the British geographical journal *Area* between Kevin Stannard and Alastair Bonnett (Stannard, 2003). The importance of linking high school and academic geography appears to emerge from this debate as a fourth issue for the survival of geography. A fifth concern, only partially recognized by any of the commentators, is the need to engage the popular media that supports the discipline, including magazines such as *National Geographic* and TV channels such as the Discovery Channel.

GIS is recognized as providing academic geography with an essential requirement for its survival: a marketable skill. This (and political savvy) will do much to ensure that AAG might be meeting in Philadelphia in 2104 to celebrate its second centenary.

At the 50th reunion of Harvard's class of 1948, veteran Foreign Service Officer Carleton Coon offered this "single pearl of wisdom" to a packed symposium on foreign policy: "Reinstitute geography at Harvard!" Now that would be something worth celebrating at any centenary.

(This column originally appeared in the December 2003 issue of GeoWorld.*)*

UCGIS Plans to Advance Geography and Improve Science

By Jerry Dobson

The University Consortium for Geographic Information Science (UCGIS) met for the first time on June 16-18, 1996, to establish research priorities in geographic information science (*www.ucgis.org*). Each member institution nominated five ideas that competed for a place on a final list. Although nominations came from individual researchers, they had to be sponsored by one or more member institutions before being presented at the UCGIS meeting. That resulted in real competition within and among institutions. Once the list was complete, an aggressive program to promote the winning priorities to federal agencies and private foundations ensued.

A Few Proposals

All UCGIS members are being encouraged to broadcast and debate their ideas on the Internet or by other means, so here were my nominations.

Formulate a new scientific method. Standard Baconian hypothesis testing hasn't changed significantly in three and a half centuries, while science has progressed in the complexity of problems it addresses and the sophistication of its technologies, methods and theories. In studying complex environments, geographers and others typically deal with at least 10 different landscape elements: spatial, temporal, cultural, processes, etc. (Dobson 1993).

While the Baconian method works well with "supporting processes and mechanisms," modern science needs spatial and process tests to address these additional elements. Such tests could help scientists who must come to grips with environmental problems having multiple geographic interactions.

Create a spatial language. Scientists digitize geographic information because there's no spatial language capable of representing spatial entities directly in a computer. Most analysis is conducted in a mathematical language, which is converted back to spatial objects whenever someone wants to view intermediate or final results. Such conversions often are cumbersome, expensive and error-prone. Egenhofer and others (1991, 1992, 1994) demonstrated that a spatial language is possible, at least for certain types of objects. That's been established further by follow-on developments and implementations.

Provide reference data for accuracy assessment. Current methods for assessing geographic data accuracy, including remotely sensed images, assume a readily available reference database. The unfortunate term "ground truth" in itself implies that a truth exists and that one can capture it by walking through it or photographing it. Yet a recent report demonstrates that's not the case (Shapiro 1995). So far, accuracy assessment research has focused mostly on the design of sampling and statistical analysis procedures. The next challenge is to provide a similar level of guidance for designing and capturing reference databases.

Develop a place model linking GIS, transport models and process models. A fundamental research challenge for geography and, indeed, for all science is to design and develop a place model that will integrate process-oriented, aspatial paradigms with spatial paradigms. A core linkage system is needed to integrate GIS functionally with a broad range of process and transport models. The result should be a unified landscape model through which many disciplines can incorporate GIS into their research without abandoning familiar models and paradigms.

Explore the continental shelf. Nowadays, it's taken for granted that the exploration age is over, at least as far as Earth is concerned. But the ocean remains less explored than the surfaces of Venus or Mars. Scientists simply don't know what's in a zone that contains, for instance, the coast that would've been exposed during the ice ages. Such exploration will require underwater video, sonar and other remote sensing technologies more sophisticated than those used in space exploration, including a 3-D GIS beyond current capabilities.

Develop a method for measuring global change. Millions of dollars have been spent on global change research. Billions have been spent on satellites and Earth sensors. When the U.S. National Oceanic and Atmospheric Administration Coastal Change Analysis Program (C-CAP) needed a method for measuring regional land cover change, however, there was little help in remote sensing and GIS literature. The situation is even less encouraging for continental or global coverage.

Generally, the remote sensing community has focused on algorithm development and experimental research for small areas. Now C-CAP has spent more than a decade developing a protocol for regional change analysis (Dobson et al. 1995). Someone needs to do the same for global change analysis.

Notice that I presented six suggestions. I couldn't whittle my list down to five, so you can imagine what it was like when 50 people from 25 institutions tried to pare a slate of more than 100 nominations down to a manageable list of research priorities.

For more information on UCGIS including research priorities and current membership go to *www.UCGIS.org*.

(This column originally appeared in the May 1996 issue of GIS World.*)*

How Can Critical GIS Be Defined?

By Jeremy Crampton

What is critical GIS, and how is it any different from "regular" GIS? And what's the relevance and importance of critical GIS?

As far as I know, the term "critical GIS" first was used in the mid-1990s. The National Council on Geographic Information and Analysis identified its 19th research initiative as that of "GIS and Society," or the social implications of using GIS. Interest in this initiative emerged from meetings held throughout the 1990s, starting in 1993 with a meeting at Friday Harbor, Wash., a lovely island community in the Puget Sound.

I was lucky enough to attend this meeting, having just returned to the United States after 18 months in the United Kingdom. Friday Harbor was something of a landmark event, because it brought together people from many different perspectives, including academics within the GIS community and those who were critical of it. Two of the outcomes of these meetings was a special issue of the journal *Cartography and GIS*, and John Pickles' book *Ground Truth* (Pickles 1995).

The questions raised at Friday Harbor included how GIS represented and affected the societies and landscapes it was serving. According to the University Consortium on GIS, critical GIS "is concerned with limitations in the ways that populations, locational conflict and natural resources are represented within current GISs, and the extent to which these limits can be overcome by extending the possibilities of geographic information technologies."

From Systems to Science

I believe that the GIS and Society meeting prompted the change from geographic information "systems" to "science," a switch that allowed the acronym GIS to be retained. Professor Michael Goodchild at the University of California, Santa Barbara is acknowledged as the progenitor of this switch (Goodchild 1992). "GI Science" can be defined as a body of knowledge and practices with social effects.

A good example is public participation GIS (PPGIS). According to one of its leading researchers, Rina Ghose at Illinois State University, PPGIS can empower marginalized communities that may have been left out of centrally made planning decisions—digital inequalities of access to information technology such as GIS or the Internet can be overcome. It's remarkable that today only about 10 percent of the world has access to the Internet, never mind GIS.

The Urban and Regional Information Systems Association hosts the annual PPGIS conference (see *www.urisa.org* for more information), where the latest research on PPGIS is presented.

Critical Is Political

This column hopes to address issues that understand GIS more critically as being embedded in and shaping society. I hope to ask difficult questions, such as those raised by fellow *GeoWorld* columnist Jerry Dobson on whether GIS contributes to "geoslavery" or invasions of privacy (Dobson and Fisher 2003).

Because critical GIS is concerned with the geographical way we understand and make decisions about ourselves, it's necessarily political. This isn't the same as saying that GIS is politicized, which usually has a negative connotation. It does mean GIS takes part in a power-laden struggle over representation. For example, take the issue of geosurveillance, or the spatial tracking of objects and people. Although there is a large technological component to geosurveillance, it also has obvious ethical (Basu 2004) and policy components (Crampton 2003). If GIS isn't political in this way, then I'm not sure what relevance it has.

(This column originally appeared in the March 2004 issue of GeoWorld.*)*

Should GIS Enter the World of Consilience?

By Nigel Waters

Many intellectual subjects seek to unite their highly disparate elements. "Grand unified mathematics" was the ultimate goal of Andrew Wiles' widely publicized proof of Fermat's Last Theorem (Singh 1998). Physicists repeatedly search for a "theory of everything." And links among scientific disciplines frequently are sought, if only to justify academic clusterings within universities.

However, few such unification attempts have had such grand, imperial designs as that of Harvard biologist Edward O. Wilson (1998a), who in his new book, *Consilience: The Unity of Knowledge*, suggests we should return to the Ionian Enchantment. This "enchantment" was the desire of Ionian philosopher Thales, who in the 6th century B.C. advocated that people unite all branches of knowledge. Wilson now suggests we unite everything under the banner of science and the scientific method, with a heavy emphasis on evolutionary biology.

The outrageousness of Wilson's polemic, that social disciplines—even the humanities—should be assimilated by the sciences (consilience literally means a jumping together), is the prime reason his book was afforded such wide attention in 1998, including discussion and reviews in *Wired* (Bennett 1998), *Scientific American* (May 1998) and even *Newsweek* (Cowley 1998, Begley 1998). In addition, Wilson's book was preceded by an extensive excerpt in *Atlantic Monthly* (Wilson 1998b).

Ironies Abound

Irony seems to surround Wilson, as indicated by the following:

Irony 1: Wilson, the scientist, has been accused of producing bad science (Begley 1998), with his insistence that evolution always produces a more fit individual. Some traits may simply be neutral byproducts. The narrowness of his stringent, neo-Darwinian viewpoint as well as the insistence on a Kipling-like "just-so" story for every trait seem bizarre.

Irony 2: The man who wrote "Biodiversity is our most valuable but least appreciated resource," in a chapter titled "Unmined Riches" in his book *The Diversity of Life* (Wilson 1992), now argues for a single methodology for all branches of learning. Thank heaven (I jest) the biological world doesn't rely on Wilson's bleak, reductionist approach to methodology. Uniformity is a bad bet in academia, the natural world or even a stock market portfolio.

Irony 3: Evolution moves faster with geographical separation. Different geographies and environments as well as physical separation should encourage greater evolution. By analogy (and don't neo-Darwinians love analogies), keeping methodologies distinct and isolated should encourage more ways of dealing with the human environment. Look at isolated Australia's fauna and flora as an example.

Irony 4: Wilson declares (1998a), "I found it hard to accept that our deepest beliefs were set in stone by agricultural societies of the eastern Mediterranean more than 2,000 years ago." Yet his deepest belief seems to concur with Thales and the Ionian Enchantment first promulgated 600 years before that.

Irony 5: A limited version of consilience isn't a "big deal" in some circles. Several universities in the United Kingdom already have degree programs in the "so-called" Human Sciences (Oxford, Sussex and University College London), with a heavy emphasis (often 50% or more) on integrating biological science and the study of human behavior with more traditional approaches such as geography, anthropology and sociology. Wilson seems unaware of this.

Irony 6: The biggest sin, according to neo-Darwinians, is to suggest that there's teleology (external guidance) in evolution, yet that's exactly what scientists introduce into genetic algorithms. However, if we have to solve ethical problems through evolutionary processes that use genetic algorithms, then we better know the goals we seek. Let's hope they're more humane than survival of the fittest!

Should GIS Move to a Science-Only Model?

GIS itself has moved down the path of scientific consilience. In 1997, the leading peer-reviewed journal in the field, *International Journal of Geographical Information Systems*, changed the last word in its title to "Science" (Fisher 1997), based on arguments made in a seminal paper by Mike Goodchild (1992). The new emphasis was to present technology as the servant—not the master—of science's traditional aims, such as data collection, measurement, analysis, modeling, theory testing, visualization and display.

Space and spatial variation, description and analysis often have been the goals of nonscientific geographers. But the extent to which they may be accommodated within GIS, let alone a newly minted Geographic Information Science, is moot. Critics such as Robert Rundstrom (1995) suggest it may be impossible to capture Traditional Environmental Knowledge (TEK), such as that held by First Nations in North America, within a GIS. It's been argued elsewhere (Clayton and Waters 1998) that perhaps new approaches and methodologies such as Case-Based Reasoning may be more sympathetic to TEK. But however this is resolved, it's likely that in geography—and even GIS—there's a need for many methodologies, rather than Wilson's consilience into a single approach.

(This column originally appeared in the November 1998 issue of GeoWorld.*)*

GIS and the Technologies of Knowing: A Proposal for GIScience?

By Nigel Waters

In his recent book, *Technologies of Knowing*, John Willinsky deplores the irrelevance and indeterminacy of much of social science research. He provocatively states that over a broad range of human activities, including bilingual education, reading instruction, welfare reform, intelligence assessment and the status of the family, the public has no confidence in the findings of social scientists.

Social science researchers themselves appear to be constantly at odds with one another. What, for example, is the best way to learn to read? Willinsky cites academics who wax scientific over the merits of phonics and, by contrast, others who grow eloquent over the wonders of the whole-language approach. As Jan de Leeuw noted (Fotheringham, 2000), social science research tends to be noncumulative. Social science researchers build molehills of knowledge and then their colleagues destroy them—rarely do they accumulate a sizable mountain.

A New Proposal

Willinsky describes the structure of a new proposal, the Automata Data Corp. (ADC), in chapter two of his book. He emphasizes that it can easily be built with existing technologies. ADC would serve as a source of information, coordination and commissioned research. Its primary presence would almost certainly be Web-based, and it would provide a multidimensional map of research contributions on any given topic in the social sciences, perhaps in the spatial format described in one of my earlier columns (Waters, 1999).

ADC would provide a public, professional forum for the discussion of completed research. Funding and grant agencies could use ADC to establish research programs that were closely linked across various populations, geographical areas and through time. ADC also would act as a brokering service to stimulate research in areas where obvious gaps had been identified.

Willinsky also details the relationship that researchers would have with ADC as well as its usefulness to the public. ADC would provide public access to information and research on social science topics, allowing the public to make more informed choices when deciding, for example, on whether or not to place their children in a bilingual education program. In addition, the public could volunteer data on themselves and their time as research subjects for future projects.

Willinsky shows that, in certain fields such as medicine, such resources already exist for professionals. At Radcliffe Hospital in Oxford, England, there's a Centre for Evidence-Based Medicine (EBM), which seeks to draw together medical research in a variety of fields (see *http://cebm.jr2.ox.ac.uk/ebmisisnt.html*). There's also a Best Evidence CD-ROM and the EBM Journal that presents peer-screened abstracts from about 70 medical journals.

Does GIS Need an ADC?

The question then arises as to whether GIS research falls into this same category of indeterminacy of results and contested methodologies. Have we not recently advocated a cut-and-dried GIScience (Goodchild et al., 1999; also see *www.giscience.org*) that has every prospect of a steady accumulation of knowledge?

In some narrowly defined, highly technical fields, this may be true. For example, since the early 1970s, algorithms that solve computational geometry problems of polygon overlay have steadily improved in speed, storage needs and size of the problems they can handle. Shortest-path algorithms in transportation-oriented GIS also have been improved consistently in all these critical areas.

By contrast, many problems to which GISs are applied commonly yield far less definitive answers. In our own research at the University of Calgary, we've used GIS to help determine the location of road mitigation strategies such as underpasses and overpasses (Alexander and Waters, in press). The goal is to help reduce the killing of wildlife such as wolves and deer along Canada's main road, the Trans-Canada Highway. There is, however, considerable debate over the effectiveness of existing mitigation structures (Savage, 2000), some arguing in support of tiny little overpasses and others for the construction of extended sections of buried or elevated highway. For politicians and decision makers to understand this research and sort out the competing claims, posting such work to an ADC Web site may make good sense.

The value of integrating "Traditional Environmental Knowledge" into a GIS also has been widely advocated. For example, in a recent contribution relating to land use and resource management, Christopher Guss (2000) argued that integrating into a GIS "First Nations knowledge ... is necessary to create and sustain a framework that will best serve generations of First Nations communities." However, it's not difficult to find strongly dissenting views concerning the value of GIS in managing aboriginal resources (Rundstrom, 1995).

Judging by its prominence at the 96th Annual Meeting of the Association of American Geographers (AAG) held April 2000 in Pittsburgh, Public Participation in GIS (PPGIS) is another controversial growth area. At the AAG conference, there were more than 20 papers devoted to PPGIS. (However, PPGIS seems to have rated only one mention on the Urban and Regional Information Systems Association's conference CD-ROM for the same year.) But PPGIS encourages conformity to the rules and procedures of GIS implementation and use, and such conformity can be problematic for activists for many reasons, including a lack of resources, the potential for co-option and ideological differences (Sieber, 2000).

Consequently, policy recommendations resulting from the implementation of PPGIS are likely to generate the same controversy and starkly different conclusions that have been produced by the type of social science research described by Willinsky. If these differences are to be resolved, or at least understood, by the general public, then an ADC for GIS research is a must.

In February 1999, the National Center for Geographic Information and Analysis (NCGIA) posted to its Varenius Web site papers that had been presented at its specialist meeting on Measuring and Representing Accessibility in the Information Age (*www.ncgia.ucsb.edu/varenius/access/*). This was a step forward, but we also need a site that will provide access to GIS-based social research. Perhaps NCGIA will take the Varenius project one step further and develop a GIS ADC? [Update: The Center for Spatially Integrated Social Science at the University of California, Santa Barbara received funding from the National Science Foundation in 2001. See *www.csiss.org*.]

What's Willinsky doing now? He has yet to persuade anyone to implement his ADC idea, but he has written another book (Willinsky, 2000; see *www.educ.ubc.ca/faculty/ctg/pkp/if.html* for Chapter 1). The book again argues for increasing the value of social science research through the use of "new formats, genres, apparatuses and technologies." Amen to that.

(This column originally appeared in the November 2000 issue of GeoWorld.*)*

Section 1. References

Alexander, S.M., Waters, N.M. 2000. "The Effects of Highway Transportation Corridors on Wildlife: A Case Study of Banff National Park," *Transportation Research Part C: Emerging Technologies,* Vol. 8, No. 1, pp. 307-320.

Basu, R. 2004. "Geosurveillance through the Mapping of Test Results: An Ethical Dilemma or Public Policy Solution?," *ACME: International E-Journal for Critical Geographies,* Vol. 3, No. 2, pp. 87-111.

Begley, S. 1998. "Why Wilson's Wrong." *Newsweek*, Vol. 131, No. 25, pp. 61-62.

Bennett, P. 1998. "From Ants to Einstein." *Wired*, Vol. 6, No. 4, pp. 178-179.

"Bluebeard on the Beach," Time. May 28, 1973, p. 31.

Boukreev, A., Dewalt, G.W. 1999. *The Climb: Tragic Ambitions on Everest*. St. Martin's Press, New York.

Clayton, D., Waters, N. 1998. "Distributed Knowledge, Distributed Processing, Distributed Users: Integrating Case-Based Reasoning and GIS for Multicriteria Decision Making." *Multicriteria Decision Making and Geographic Information Systems*. New York.

Cowley, G. 1998. "Wilson's World." *Newsweek*, Vol. 131, No. 25, pp. 58-61.

Crampton, J.W. 2003. "Cartographic Rationality and the Politics of Geosurveillance and Security." *Cartography and Geographic Information Systems*. Vol. 30, No. 2, pp. 131-44.

Dobson, J.E. 1991. "Lack of Spatial Intelligence Hampers Communication," GIS World, Vol. 4, No. 7.

Dobson, J.E. 1993. "The Geographic Revolution: A Retrospective on the Age of Automated Geography," *The Professional Geographer*, Vol. 45, No. 4.

Dobson, J.E. and Fisher, P.F. 2003. "Geoslavery." *IEEE Technology and Society Magazine*, Vol. 22, No. 1, pp. 47-52.

Dobson, J.E., 1998. "Is GIS a Privacy Threat?," GIS World, Vol. 11, No. 7, p. 34.

Dobson, J.E., Bright, E.A., Ferguson, R.L., Field, D.W., Wood, L.L., Haddad, K.D., Iredale III, H., Klemas, V.V., Orth, R.J., Thomas, J.P. 1995. NOAA Coastal Change Analysis Program, Guidance for Regional Implementation, Version 1.0, NOAA Technical Report NMFS 123.

Egenhofer, M.J. and J.R. Herring. 1992. *Categorizing Binary Topological Relationships Between Regions, Lines and Points in Geographic Databases*, unpublished manuscript.

Egenhofer, M.J., Franzosa, R. 1991. "Point-Set Topological Spatial Relations," *International Journal of Geographical Information Systems,* Vol. 5, No. 2, pp. 161-174.

Egenhofer, M.J., Mark, D., Herring, J.R. 1994. "The 9-Intersection: Formalism and Its Use for Natural-Language Spatial Predicates," Technical Paper 94-1, National Center for Geographic Information and Analysis, Santa Barbara, Calif.

Einstein, A. 1954. "The Cause of the Formation of Meanders in the Courses of Rivers and of the So-called Baer's Law," *Ideas and Opinions*. Bonanza Books, New York, pp. 249-253.

Finney, B. 1994. *Voyage of Rediscovery: A Cultural Odyssey through Polynesia*. University of California Press, Berkeley, Calif.

Fisher, P. 1997. "Editorial." *International Journal of Geographical Information Science*, Vol. 11, pp. 1-3.

Fotheringham, A.S., Brunsdon, C., Charlton, M. 2000. *Quantitative Geography*. Sage Publications, London.

Gaiel, G.L., Willmott, C.J. (eds.). 2003. *Geography in America at the Dawn of the 21st Century*. Oxford University Press, New York.

Ghose, R. "Use of Information Technology for Community Empowerment: Transforming Geographic Information Systems into Community Information Systems," *Transactions in GIS*, Vol. 5, No. 2, pp. 141-163.

Goodchild, M.F., Egenhofer, M.J., Kemp, K.K., Mark, D.M., Sheppard, E. 1999. "Introduction to the Varenius Project," *International Journal of Geographical Information Science*, Vol. 13, pp. 731-745.

Goodchild, M.F. 1992. "Geographical Information Science," *International Journal of Geographical Information Systems,* Vol. 6, No. 1, pp. 31-45.

Guss, C. 2000. "Wisdom of Digital Information Management and British Columbia's First Nations," CD-ROM of Proceedings, GIS 2000 Conference, GeoTec Media, Boulder, Colo.

Harris, T.M., Weiner, D. 1996. "GIS and Society: The Social Implications of How People, Space, and Environment are Represented in GIS," *Technical Report 96-7*, National Center for Geographic Information and Analysis, Santa Barbara, Calif.

Jalbert, J.S., Dobson, J.E. 1977. *A Cell-Based Land Use Screening Procedure for Regional Siting Analysis*, ORNL-NUREG/TM-80, Oak Ridge National Laboratory.

Johnston, R. 2002. "Reflections on Nigel Thrift's Optimism: Political Strategies to Implement His Vision," *Geoforum*, Vol. 33, pp. 421-425.

Krakauer, J. 1998. *Into Thin Air: A Personal Account of the Mount Everest Disaster*. Villard Books, New York.

Larson, E. 2000. *Isaac's Storm*. Vintage Books, New York.

Martin, G.M. 1998. "The Emergence and Development of Geographic Thought in New England," *Economic Geography*, Vol. 74, pp. 1-13.

May, R.M. 1998. "An Attempt to Link the Sciences and the Humanities." *Scientific American*, Vol. 278, No. 6, pp. 97-98.

Pattison, W. D. 1964. "The Four Traditions of Geography." *The Journal of Geography*. Vol. 63, pp. 211-216.

Philbrick, N. 2000. *In the Heart of the Sea*. Viking, New York.

Pickles, J. 1995. *Ground Truth*. Guilford, New York.

Rundstrom, R.A. 1995. "GIS, Indigenous Peoples, and Epistemological Diversity." *Cartography and Geographic Information Systems*, Vol. 22, pp. 45-57.

Rundstrom, R.A. 1995. "GIS, Indigenous Peoples, Epistemological Diversity," *Cartography and Geographic Information Systems*, Vol. 22, pp. 45-57.

Savage, C. 2000. "A Highway Runs Through It: Underpasses, Overpasses and Fencing Protect Banff's Wildlife from Traffic on the Trans-Canada Highway. Are They Enough?" *Canadian Geographic*, Vol. 120, No. 5, pp. 34-42.

Shapiro, C. 1995. "Coordination and Integration of Wetland Data for Status and Trends and Inventory Estimates," Federal Geographic Data Committee, Wetlands Subcommittee, Technical Report Number 2.

Sieber, R. 2000. "Conforming (to) the Opposition—Geographic Information Systems in the Conservation Movement," CD-ROM of Abstracts, AAG Annual Conference, AAG, Washington, D.C.

Singh, S. 1998. *Fermat's Last Theorem*. Fourth Estate, London.

Smith, N. 1987. "Academic War Over the Field of Geography: The Elimination of Geography at Harvard, 1947-1951," *Annals of the Association of American Geographers*, Vol. 77, pp. 155-172.

Stannard, K. 2003. "Earth to Academia: On the Need to Reconnect University and School Geography," *Area*, Vol. 35, pp. 316-322.

Stone, I. 1956. *Men to Match My Mountains*. Doubleday & Co. Inc., New York.

Tenner, E. 1996. *Why Things Bite Back: Technology and the Revenge of Unintended*

Consequences. Knopf, New York.

Thrift, N. 2002. "The Future of Geography," *Geoforum*, Vol. 33, pp. 291-298.

Tisma, A. 2000. "Information Design for Public Decision Making: Case of the Master Plan for the South Axis in Amsterdam," CD-ROM of URISA 2000 Proceedings, URISA, Park Ridge, Ill.

Waters, N. 1999. "New Software Organize Information Spatially," *GeoWorld*, Vol. 12, No. 5, p. 34

Waters, N. 2003. "Traffic Safety Rcmains an Ongoing Concern," *GeoWorld*, Vol. 16, No. 6, p. 22.

Wilford, J.N. 1981 (revised 2000). *The Mapmakers*. Alfred A. Knopf, New York.

Willinsky, J. 1999. *Technologies of Knowing: A Proposal for the Human Sciences*. Beacon Press, Boston.

Willinsky, J. 2000. *If Only We Knew: Increasing the Public Value of Social Science Research*. Routledge, New York.

Wilson, E.O. 1998. *Consilience: The Unity of Knowledge*. Knopf, New York.

Wilson, E.O. 1998. "Back from Chaos." *The Atlantic Monthly*, Vol. 281, No. 3, pp. 41-62.

Wilson, E.O. 1992. *The Diversity of Life*. Harvard University Press, Cambridge, Mass.

Worsley, F.A. 1977 (reissued 1998). *Shackleton's Boat Journey*. W.W. Norton & Co. Inc., New York.

Section 2.
GIS, History and Politics

Introduction
By Nigel Waters

The second section of this book is concerned with the origins of GIS, its history and its effect on history as well as how it has and should play a role in the political events of the day.

In the opening two articles, Jeremy Crampton brings together the origins of the disciplines of geography and GIS. In the first article, he shows that we have always been doing GIS-like operations, but now it's easier, because we have new technology.

Crampton explores the various disciplinary routes from the advent of the first AAG Meeting in Philadelphia in 1904. He's able to show the link between Davis' cycle of erosion and the modern methodologies employed by the "father of GIS," Roger Tomlinson. Links between Emory Johnson's Wharton School at Pennsylvania and today's business geographics also are explored. Early research in population geography is shown to have been essential to the study of today's GIS-enabled population geography—the foundation of many urban modeling studies.

Crampton's second piece reviews the work of Tom Patterson, a leading landscape cartographer at the U.S. National Parks Service. Crampton shows that technology can be viewed as an art—as the Greek origins of the word suggest—and provides information on how to access some of Patterson's "awe-inspiring" cartographic images.

Jerry Dobson also is concerned with the origins of GIS in his discussion of Tim Foresman's Geomander, a visual representation of the key events in the history of GIS. Such a diagram will never include every development, and thus won't satisfy everyone. Dobson notes some of these omissions and, in doing so, draws attention to the work of William Warntz.

"In that fateful year of 1975, when the academic job market went into a tailspin (just after I accepted my first academic appointment), I co-authored my first peer-reviewed publication with William Warntz and have fond memories of the help that he provided to my fledgling career," writes Dobson.

In his third contribution, Crampton provides an insider's look at the monumental work on the *History of Cartography*. He's on the board of advisers for the sixth volume in this landmark series.

GIS is used for military purposes, but, as Crampton shows in his next article, maps (and by extension GIS) can be used to build peace. This was the hope at the Paris Peace Conference that opened on Jan. 12, 1919, after "the war to end all wars." Crampton provides a fascinating look at the role that American geographers played in the peace process.

Counterfactual history is a thought-provoking and well-regarded academic subdiscipline, at least in some circles. Robert Fogel won the 1993 Nobel Prize for Economics, in part, for his counter-factual exploration of American economic history without the railroad. The sixth contribution in this section is a light-hearted examination of how influential "perfect spatial knowledge" might have been for some leading historical figures had they had access to "GIS-type" technology.

Spatial ignorance and its effect on foreign policy is the theme of Dobson's second item (our seventh selection). It's depressing to realize that we're repeating the mistakes of the past, and that, in some ways, our ignorance of the real world is increasing.

It's also depressing to think that we've learned so little since Sir John Mandeville wrote his plea for tolerance and understanding of other civilizations and religions from his 14th century English home (Milton, 1996). Perhaps we do need the new, macro political geography discussed later in the 10th article in this section, and perhaps GIS can do a little to avoid a clash of civilizations (Huntington, 1998).

Dobson's gem (our eighth selection) on the Von Thunen Museum at Tellow also pleas for information over ignorance. It's an intriguing story of how one individual rescued material pertaining to one of the most famous models in economic geography. Readers wanting more information might want to look at the Web site of the North American branch of the von Thunen society at *www.personal.umich.edu/~nystuen/thunen/thunen*.

We revisit the theme of spatial ignorance in the tale of the horror that King Leopold II of Belgium visited on the inhabitants of the Congo in the late 19th century. Dobson's tale is timely, because the horror continues more than 100 years later.

The last two articles I wrote to highlight the value of making data available for free as well as to demonstrate the importance of a broad social-science perspective in designing GIS and using GIS software. Although the first article was written almost six years ago, the cost-recovery policies of federal and municipal governments, such as the city of Calgary, continue to damage, limit and restrict development of the GIS industry in Canada. I have written again on this topic in my February 2006 *GeoWorld* column (*www.geoplace.com/uploads/FeatureArticle/0602en.asp*).

Such policies continue to damage the prospects for the future of GIS so dramatically envisaged in Jack Dangermond's Foreword to this book. The second and last articles in this section argue for a broad view of the role of GIS. Yes, the "G" in GIS is important. Geography is vital, but embracing the contributions of other disciplines also is critical.

Have We Seen 100 Years of GIS?

By Jeremy Crampton

I believe we are celebrating 100 years of GIS. But how can this be, when most histories date GIS to the late 1960s?

The answer is that in March 2004, the Association of American Geographers (AAG) will hold its centennial meeting in Philadelphia. And the history of AAG also is the history of GIS. Many of the ideas behind GIS were developed by geographers and presented at AAG meetings, discussed in its journals or influenced by leaders in AAG and related organizations.

AAG was founded on Dec. 29, 1904, in Room 16 of College Hall at the University of Pennsylvania in Philadelphia. William Morris Davis, the famous Harvard geomorphologist, was the first president.

Looking at some of the problems tackled by these early geographers, we see that their questions are still pursued in GIS today. What's more, many of the map types and quantitative methods now used in GIS were invented in the early 20th century.

Resource Management

The methods developed long ago include resource management as well as political and economic geography. Davis himself took part in a survey of Montana to assess and map its resources that might provide revenue for the Northern Pacific Railroad. Such efforts to manage and evaluate land now are primary functions of GIS.

Davis authored the "cycle of erosion," which is the idea that landforms evolve through a series of uplifts, erosion and further uplifts. Prior to his theory, many geologists believed that land was static and changeless. Roger Tomlinson, who invented the Canadian GIS in the 1960s, applied the concepts to forest management.

Transportation Planning

Meanwhile, at the University of Pennsylvania's Wharton School, a strong tradition of economic geography was emerging. One of its subjects was transportation planning. Emory Johnson, dean of the Wharton School, was instrumental in passing the landmark Transportation Act of 1920, which returned railroad ownership to private hands following its appropriation by President Woodrow Wilson for the war effort.

Johnson also trained many of the nation's economic geographers. In 1901, Johnson opened a "Geography and Commerce" division in his department. Today, business applications are an important part of GIS.

Population Mapping

Perhaps the most fascinating "proto-GIS" work was in population studies. For the cartographers and geographers involved in cartography, the mapping of populations and their attributes such as ethnicity, language, origin and density was a critical issue. These scholars saw the problem as a statistical one of population management.

By the early 1930s, for example, geographer J.K. Wright was so taken up with the issue of "the cartographic presentation of statistical data and the devising of summary measures of distributions on statistical maps" that it became known to his wife and children as "The Problem" (Bowden, 1970).

Population mapping remains a major concern. My *GeoWorld* colleague Jerry Dobson, for example, has been instrumental in developing a better assessment of global population distributions through his work on LandScan at the Oak Ridge National Laboratory (see Plate 1).

LandScan addresses a problem faced by census enumerators: collecting reliable data. Many countries can't afford to do so or can't do so reliably for the whole country. The solution adopted in LandScan is to supplement census data with remote-sensing imagery. Theoretically, according to Dobson, the Department of Defense-funded project can provide resolutions down to the city block.

Of course, much of this work predates the invention of GIS as a technology by many decades and wasn't directed specifically toward what would become GIS. But nonetheless, it later proved to be essential to the development of GIS. Without it, GIS may not have developed, and it certainly wouldn't look the way it does today.

(This column first appeared in the January 2004 issue of GeoWorld.*)*

Technology Returns to Its Origins

By Jeremy Crampton

Cartographic representation of a landscape has always posed a special challenge in mapmaking and GIS. Unlike a thematic or statistical map, the challenge lies in evoking a sense of place—some essence or even spirit of the landscape. The drive that's occurring in GIS toward realistic 3-D representations of the landscape would seem to be the industry's finest technological achievement toward this goal.

However, this is only true if we think of "technological" in its original sense. The Greek *Technê* means "arts" or "skill." An artisan is a skilled craftsman who produces something. The modern meaning of technology has evolved, but, originally, technology was an art.

Practicing Art

With this in mind, we can properly appreciate the work of people who manage to infuse technology and art in their depictions of landscape. One such practitioner is Tom Patterson, a leading cartographer at the National Park Service. For several years, Patterson has been working with new technologies in the tradition of great cartographers such as Heinrich Berann and Eduard Imhof to evoke and reveal landscapes.

One of Patterson's secrets—besides having technical mastery of his digital tools—is that he first visits and absorbs the places he will reproduce in his maps. Through this dedication to detail and commitment, Patterson has produced some beautiful landscapes while maintaining accuracy and scientific precision.

More than a Photo

Oregon's Crater Lake is one of his latest works. In this map (Plate 2), we see how he can produce a "sense" of the crater—not like a photograph, because we can see the underwater rises and islands as well as the roadway around the edge, but rather as a completely realized landscape.

Author's Note: For more examples and explanations of Tom Patterson's techniques, visit his Web site at *www.shadedrelief.com.*

(This column originally appeared in the July 2003 issue of GeoWorld.*)*

Foresman's Geomander Shows GIS Origins

By Jerry Dobson

For anyone interested in the history of GIS, there's an indispensable book featuring a complex diagram that reveals a lot about how GIS came to be. I'm referring to *The History of Geographic Information Systems: Perspectives from the Pioneers*, edited by Tim Foresman (1998), and his "historical geomander" of the "genesis of modern GIS."

The Geomander

Foresman's geomander is designed so the earliest developments appear at the periphery. Embryonic events in the outer tier go back to Herman Hollerith's invention of the famous card that bears his name and the resulting automation that allowed the 1890 Census to be tabulated before the 1880 Census and led eventually to Hollerith's founding of IBM. Seventy years later, the U.S. Census Bureau stands out again with its development of topological data structures. Don Cooke's work for the Census Bureau and Roger Tomlinson's work for Environment Canada, prominent in the diagram, are documented in fascinating detail in chapters written by the two pioneers.

The diagram is complex, because so many different organizations contributed to the birth and maturation of GIS. To rate a place in the outer tier (i.e., those present from conception through delivery), organizations must have been engaged in GIS at least as early as the 1970s. Eighteen organizations are listed, notably including the Harvard Laboratory for Computer Graphics and the groundbreaking work there by Howard Fisher and Carl Steinitz.

The geomander's central wheel represents the formative years of GIS from the 1980s through the 1990s. Thirty-seven organizations lie on spokes that touch the hub, including *GeoWorld* (then GIS World).

Most favored of all are the organizations with long spokes touching the hub, yet reaching far back. The longevity prize goes to the Census Bureau, with the earliest events culminating in Topologically Integrated Geographic Encoding and Referencing (TIGER) at the hub. The U.S. Geological Survey (USGS), Urban and Regional Information Systems Association and several prominent GIS vendors round out the list.

Omissions?

Foresman told me he's heard some grousing about the geomander, but I find it to be a good representation compared to other accounts. I found only three items that seemed to be missing.

Here and in a separate textbook, Nick Chrisman emphasized the importance of John K. Wright's early advancements in mathematical cartography, especially dasymetric interpolation, as a foundation of GIS. Chrisman also emphasized William Warntz's developments in "social physics" as a foundation (not yet fully realized) for representing social and cultural phenomena in a quantitative fashion. Richard Durfee and I emphasized the importance of early standards efforts. The National Committee for Digital Cartographic Data Standards, in particular, deserves a line on the geomander.

Intertwined Roots

When the geomander first appeared, I noted with pride that my institution at the time, Oak Ridge National Laboratory (ORNL), appears in one corner (under the rubric of the Department of Energy). Later, I noticed that the Oak Ridge Regional Modeling Information System (ORRMIS), the first of several names we used for the Laboratory's GIS capability, reaches from there to the hub. That placed us in the favored group of long-term contributors, and I felt more than a little pride.

Then last semester I taught a course on the historical development of geographic thought, methods and technology, and that forced me to look more closely. I was surprised at how intertwined the roots of GIS turned out to be. I should have known already, because the same people kept showing up at meetings in those early years, often wearing different hats to represent their employer, sponsor or professional organization. They still do, for that matter.

Memorable Connections

My first GIS sighting was at USGS, where I saw a demonstration during a visit in 1966. I took my first GIS course in 1973 under Jim Carter at the University of Tennessee, which doesn't show up in the geomander. We used SYMAP, which does show up, but my connection to it was too distant for bragging rights.

In the final year of my Ph.D. program, the academic job market for geographers hit the bottom. With 75 rejection letters in hand, I told myself not to worry, but rather concentrate on my dissertation. When the graduation ceremony was over in March 1975, I told myself it was indeed time to worry. That afternoon, ORNL called with an offer. As it happened, I was the proverbial "unemployed Ph.D." for about five hours.

Only after I joined the staff did I realize just how advanced ORNL was in GIS. Development had begun there in 1969, making ORNL one of the first GIS centers in the world, and ORRMIS was the first GIS designed for regional modeling. In an earlier column, I mentioned the rich flow of people and ideas, but I didn't realize until now how prominent many of those connections turned out to be.

My first assignment was an ORNL project to design and implement a power-plant siting methodology for the state of Maryland based on the Maryland Automated

Geographic Information (MAGI) system, which shows up in the venerable middle tier of Foresman's diagram. MAGI was being developed by Jack Dangermond for the Maryland Office of Planning, led at that time by John Antenucci, who later founded Plan-Graphics. My Geographical Review article reporting the results of our study is one of the earliest journal articles demonstrating the analytical potential of GIS (Dobson, 1979).

In 1983, I proposed and led the effort to fund the National Center for Geographic Information and Analysis (NCGIA). Chrisman calls NCGIA the "seal of approval" of geographic information science by the National Science Foundation and the scientific community. Later, I helped John Bossler, Will Craig, Michael Goodchild, Karen Kemp, David Mark and others found the University Consortium for Geographic Information Science and served as its third president. Starting with 30 core institutions, UCGIS later grew to 62 universities, four professional associations, six private corporations, two government agencies and two international affiliates.

In addition, I helped found the Association of American Geographers' GIS Specialty Group and served in various offices, including chair, from 1987 through 1992. I've been a contributing editor with this column in GeoWorld since 1991. I consulted briefly to ERDAS when it was a small organization and visited Comarc during its short span in San Francisco. Durfee and I visited Bill Goran when GRASS was just a sprig. I visited Dangermond at ESRI just months before the appearance of PC ARC/INFO and spoke at the Intergraph Users' Conference the year TIGRIS was announced.

Now, I'm president of the American Geographical Society, where Wright worked when he invented dasymetric interpolation and Warntz worked when he started with social physics. And I've joined the faculty of the University of Kansas, which Foresman recognized in his most venerable outer tier.

In retrospect, I'm startled to realize that I personally participated in six separate lines on the geomander, collaborated or visited with seven other key organizations in their formative years, and later affiliated with two other organizations listed by Foresman and/or Chrisman. My personal recollections are every bit as convoluted and intertwined as Foresman's geomander implies.

(This column originally appeared in the March 2003 issue of GeoWorld.*)*

Knowing the Past Can Provide Focus for the Future

By Jeremy Crampton

Many people in the GIS and mapping industry originally entered it because of an early love of maps. When I was a kid, I was fascinated with maps, especially those in my favorite children's books such as *Treasure Island* (the first map to appear in a children's book), *The Lord of the Rings* and *Swallows and Amazons*.

Arthur Ransome's *Swallows and Amazons* was particularly imaginative, because Ransome remapped and relabeled the everyday places of Lake Windermere with the exotic towns and cities of South America. Perhaps you have your own childhood maps you fondly remember, and, if so, I'd like to hear about them at my e-mail address below.

Many in the industry also love old maps such as the unusual maps of the Middle Ages, which are based on a combination of religious worldviews and actual place names. I often think of the Hereford mappa mundi, a "T-in-O" map that combines biblical scenes (Noah's Ark, Moses) with contemporary geographical locations (Paris and Jerusalem). Also intriguing are maps of the Age of Discovery, when European maps filled in "the New World" (although the Americas had human occupation for at least 10,000 years), or seeing California mapped as an island long after explorers showed it to be a peninsula.

Historical Excitement

Such lovers of cartographic history need look no further than *The History of Cartography* series, an ongoing project conceived by David Woodward and Brian Harley that attempts to recapture the excitement of such historical maps.

In May 1977, Woodward and Harley fleshed out an idea for a history of cartography that would be thoroughly modern, based on new scholarship and written to the highest standards. Designed to do more than just update previous histories, the project would redefine maps and mapping.

Their task wouldn't be merely to tell the story of maps, but to describe the triumph of knowledge and science over unreason, religious dogma and superstition. With Harley as co-editor, the project would become a critical history in which claims to objectivity and naturalism would be examined and, perhaps more radically, the

Western definition of a map expanded to include indigenous and non-Western mapping traditions (some brought to light in the West for the first time).

Although not a predominant theme, the series wouldn't shy away from the "power politics" exercised in mapping territories as well as the knowledge produced in doing so. Maps were to be understood not just as efficient documents recording the truth of the landscape, but as active instruments in producing the truth.

A Massive Project

The project's scope quickly got away from them. Originally, three volumes were conceived, with the history terminating in 1900. But since 1977, it has grown to six volumes (and Volume Two is actually three books) stretching through the 20th century. Volume Three, which has taken more than five years to produce, includes 1.3 million words and more than 60 contributors, covering the years 1470-1640.

The accompanying figure shows Woodward next to the stack of papers for Volume Three, which is scheduled for publication in 2005. Volumes Four and Five take us through 1900. The Sixth volume, co-edited by Mark Monmonier, will cover the last 100 years or so, including the emergence of GIS, Internet cartography and the commercial GIS industry (I'm on the board of advisors for this volume).

Critical Approach

Besides *The History of Cartography*, Harley (who died in 1991) is known for his critical approach to mapping: being aware and respectful of the effects of your work. But I often find that those in commercial GIS already are "critical" and thoughtful about such effects—it's just that the job allows little scope to express this. Deadlines, bosses and the demands of the GIS industry all combine to divorce our work from thinking about its effects, either positive or negative.

The History of Cartography is one of the most significant intellectual endeavours in GIS and mapping. It's reawakening excitement about where we've come from, and, because of that, it can indicate where we might want to go next.

(This column originally appeared in the October 2004 issue of GeoWorld.*)*

Can We Learn from the Paris Peace Conference of 1919?

By Jeremy Crampton

The use of maps and aerial imagery for military purposes is well known, but what about when the fighting stops? This is a relevant topic at the moment as the story of GIS, Global Positioning System (GPS) technology and digital maps is told daily on the news in places such as Iraq or Afghanistan.

An earlier peacetime use of maps occurred following World War I. The "war to end all wars" was huge, with perhaps more than 10 million killed in combat. It saw the end of three empires: the Ottoman, the Austro-Hungarian and the Russian. New methods of fighting were introduced, including airplanes, tanks and submarines. But while it was still going on, experts were being brought together in several countries to decide what would happen when it was all over, and maps were a central part of the effort.

The Inquiry

After the United States joined the war in 1917, President Wilson established a secret research group made up of historians, economists, political scientists, boundary experts, lawyers and cartographers. In fact, the "Inquiry" was headquartered in the American Geographic Society (AGS) New York offices and was led by Isaiah Bowman, AGS director.

Bowman became chief territorial specialist, and he quickly tapped his former geography professor, Mark Jefferson, to be chief cartographer. At its height, the Inquiry consisted of about 150 men and women. Several famous names in geography were involved, including Ellen Churchill Semple, Eugene van Cleef (a student of John Paul Goode of the *Goode's Atlas*), Walter Briesemeister, Armin Lobeck and Douglas Johnson.

The Inquiry believed that if it could acquire enough information about each country, then it would be in a position to provide unbiased advice to policy makers. For this purpose, those involved acquired hundreds of maps and drafted hundreds of their own. Then on Dec. 4, 1918, just one month after hostilities ceased, a select group from the Inquiry boarded the USS George Washington. President Wilson and his wife soon joined them, and, to great fanfare, they set sail for France, where the peace talks would take place.

Agendas Set

Once onboard, President Wilson gave the Inquiry a general briefing. AGS Director Bowman wrote down the only known transcript of it, including an oft-repeated quote from the president: "give me a firm position, and I'll fight for it!"

Jefferson soon was busy drafting maps for the talks, and he set up a map committee with his British and French counterparts. Luckily, historian Geoffrey Martin recovered Jefferson's daily diary. Jefferson offers some acerbic asides in the diary, such as this entry for Feb. 10, 1919: "Jonesey ... wanted a map of German as probable on 3 millionth [1:3m] all changes on ample legend. Haskins said some things about Schleswig, Belgium and Alsace [areas of disputed borders] mostly to show his problems were different. As if not all the same one: to be or not to be German?"

That was the question. With the defeat of Germany, its territorial boundaries were redrawn based on maps of linguistic regions, ethnic enclaves and historical precedence. The trouble was, everybody had arguments that favored their own side.

Success or Failure?

So where would the new borders go? The treaty, which eventually was handed to the Germans in June 1919, often has been credited with establishing the geography of Europe throughout the rest of the 20th Century (see Plate 3). Just think of the Balkans—or the Middle East. In the 1930s, however, Hitler would use the tough provisions on Germany to further his own plans. Obviously, it didn't prevent another war within a generation, so did it fail? And did the maps fail to solve the problem of territory?

This is an open question. I'd suggest, however, that by emphasizing how its maps were objective and scientific, while everyone else's were partisan and biased, that the Inquiry failed to grasp what should have been obvious: all maps are political in some way.

The idea of purely disinterested scientific maps needs to be reexamined. Instead of working against politics in maps, if we fully understand them, then politics can be given a useful role.

(This column originally appeared in the October 2003 issue of GeoWorld.*)*

GIS Courts Counterfactual History

By Nigel Waters

During the summer of 1995, the British scientific journal *New Scientist* ran a competition in which entrants had to describe a famous historical event that would have turned out differently if the participants had the benefits of modern scientific or technical know-how.

Counterfactual analysis has a long, established track record in the social sciences. Robert Fogel shared the 1993 Nobel Prize for economics largely because of his analysis of a 19th-century American economy without railroads (Fogel 1964). Unlike Fogel, *New Scientist* was interested in how things would have changed if a technology had come sooner, rather than later, than its true introduction date.

My Kingdom for a GIS

Not wishing to anticipate or forestall anyone's creativity, I'd like to suggest that the fate of Richard III, king of England, might have been different if he'd had GIS and Global Positioning System (GPS) technologies. If you remember your English history, Richard III was last seen fighting for his life at the Battle of Bosworth Field in 1485. His final words in Shakespeare's play of the same name are: "A horse! A horse! My kingdom for a horse!" In that scene he appears to want to escape, regroup and fight another day—although a few lines earlier he declined Catesby's offer to look for a horse. Moments later he lies dead on the battlefield, having been dispatched by Henry, duke of Richmond (Hammond 1981).

Now suppose a horse had been found using GPS technology, and suppose the horse had an on-board navigation system. Then Richard III might have escaped, especially if the GIS stored within the on-board computer included a digital elevation model (DEM) that showed a convenient escape route.

Thus, Richard might have survived and defeated Richmond (the future Henry VII and first of the Tudors) at a later date. If Richmond hadn't won, he wouldn't have married the late Richard III's intended wife, Elizabeth of York, and then there would have been no Henry VIII. There would have been no Tudor England, no Reformation and perhaps no English exploration of the Americas (Henry VII supported the Cabots, and his granddaughter, Elizabeth I, aided seafarers such as Drake and Raleigh). North America may have evolved as a French or Spanish colony—or perhaps a mixture of both. The Tudor monarchs, Sir Thomas Moore and Shakespeare—all of whom blackened the character of Richard III—also created the myth he was deformed (Hammond 1993). Linking deformity with mendacity and an evil nature gripped by superstitious minds of the population for centuries, with unfortunate results for those with such deformities.

Custer's Great Victory

When I asked my introductory GIS class to provide me with instances of how the technology might have changed history, one graduate student in archaeology, Armando Anaya, noted that technology doesn't always have beneficial outcomes. He argued that if General George Custer had spent some money hiring a GIS consultant, he might have been able to estimate the resources available to the Sioux and their allies at the Little Big Horn encampment. That might have led to more reliable estimates of the camp's population and its age and gender breakdown, i.e., an estimate of the number of warriors. Again a DEM would have come in handy, providing a more auspicious line of attack and coordinating cavalry columns. Custer would have emerged victorious, with even more disastrous consequences for the native American population.

In Anaya's scenario, Custer is a military hero and goes on to win the next U.S. presidential election. Frustrated with the protection afforded to the remnants of Sitting Bull's forces in Canadian territory, President Custer pursues an expansionist policy with his neighbor to the North, unilaterally annexing Canada's Northwest Territories and the West Coast settlements. In the 1990s, the small, Anglophone minority of Canada secedes from the predominantly Francophone state of Canada.

In fact, Custer has spawned at least one counterfactual novel by Douglas Jones (1976), who begins his story with the following statement: "This is a fantasy which needs no apology, for who among us has not been intrigued by the alternatives history never reveals?"

The consequences of an early GIS introduction can be good or bad, but the outcomes can result in a dramatic or amusing change of history. Now, can you think of any other famous historical events that would have turned out differently with the benefits of GIS and related technologies?

(This column originally appeared in the January 1996 issue of GIS World.*)*

Geographic Ignorance Drives Foreign Policy

By Jerry Dobson

If I could choose a single fact to illustrate the lofty heights to which geography has risen in previous times, it would be the following: The American Geographical Society (AGS) was responsible for drafting President Woodrow Wilson's "Fourteen Points," certainly one of the most important and best-known policy statements of the previous century (Smith, 2003).

However, that momentous event has been so thoroughly forgotten by society and even by geographers themselves that I, current AGS president, didn't know about it until a few weeks ago. But what do these startling facts reveal about the state of geography's past, present and future?

From Lofty Heights

On the eve of World War I, Wilson knew the United States was poised to become a world power. The Great War and subsequent peace would be his nation's debut on the global stage. He relished the role of leading man, but who would play his supporting cast?

The scholar in him realized that his own government lacked the requisite expertise, knowledge and experience. For 140 years, America had practiced isolationism from world affairs, and no one in government—not even the officers and analysts of the State Department or Military Intelligence—was ready to face sophisticated European negotiators.

The problem was geographic ignorance, and Wilson turned to AGS for help. He did so tentatively at first, placing top aides in charge. Soon, however, he relied almost exclusively on AGS Director Isaiah Bowman to lead "The Inquiry's" 150 scholars who represented geography and many other disciplines. Their task was to collect and analyze the information that would be needed to establish a "scientific" peace at the war's end.

When Wilson and the American delegation left for Paris aboard the USS George Washington, Bowman sailed with them. Initially, he was supposed to report to the commissioners through official channels in the Department of State. On arrival, however, Bowman pulled off a bureaucratic coup, and Wilson decreed that analysts from the Department of State, Military Intelligence and Central Bureau of Statistics all would report through Bowman.

By late January 1919, AGS geographers and cartographers, led by Mark Jefferson, turned out more than 300 maps per week based on geographic analysis of the Inquiry's massive data collections covering language, ethnicity, resources, historic boundaries and other pertinent information. America's delegation became the envy of Versailles.

To Abysmal Depths

Now compare 2003 with 1917. The common thread is geographic ignorance. Wilson recognized it; modern politicians do not.

America's foreign policy suffers from profound geographic ignorance about the rest of the world, especially the Muslim world. Many people are fond of pointing out the president's gaffs or the opposing party's gaffs, but no one holds a corner on the market for geographic ignorance—it's bipartisan; pervasive among politicians, analysts, journalists and voters alike; and it's worldwide.

Hardly anyone in the West truly understands the East and vice versa. They're as ignorant of us as we are of them, largely due to exaggerated images broadcast by our own popular culture. Ignorance of allies and adversaries drives the foreign policies of all nations, not just America. Sadly, geographic ignorance today is the principal determinant of international relations, and no administration since World War II has known where to turn for answers.

Imagine for a moment how different U.S. policy might have been in Afghanistan and Iraq if a broad segment of society had understood the same factors for these regions that AGS geographers and cartographers mapped for Europe in 1919.

Every President Needs His Geographer

The Chou Emperor of ancient China had his geographer-royal. Louis XV had his geographer. The Continental Congress appointed its official geographer. Woodrow Wilson and Franklin D. Roosevelt had their geographer (Bowman). Since World War II, however, no U.S. president has had a geographer. (It should be noted that President Jimmy Carter's son, Jeff, is a professional geographer, and he influenced the White House to install its first GIS, the short-lived Decision Information Display System, which focused on domestic political strategy.)

Is it any wonder that we're so prone to quagmires of the Vietnam variety? As early as 1949, Bowman warned that "we can lose our shirts in the swamps and canyons of the hinterlands" of what was then known as Indo-China.

It's tempting to bemoan the loss of influence Bowman had in international affairs, but geographers still advise at high levels on similar matters. Bill Wood, deputy assistant secretary and director of the Office of the Geographer and Global Issues in the Department of State is a prime example. They're conscientious, intelligent and informed, but they stand alone.

What's lost is the mutually beneficial dialog that previously existed between national leaders and geographers outside of government. We, who once served as allies, now are supplicants in a competitive bidding arrangement that's inherently adversarial. There's no feedback mechanism whereby geography's successes in government can be translated into benefits that help move the academic and professional communities forward.

We can't be responsive to the needs of government officials, even to educate the labor force they need. And we're not at their side when they take controversial stands, as with the State Department's recently reported, pre-engagement forewarning about subsequent problems in Iraq.

A New Heyday?

From the 1400s through about 1950, geography thrived due to exploration. From the 1600s to about 1900, geography thrived due to westward expansion. In the first half of the 20th century, Bowman's heyday, geography thrived due to territorial issues in international affairs. But after World War II, there was consensus that national boundaries had been set. Politicians and colleagues in peer disciplines strangely interpreted this to mean that geography itself no longer mattered.

Surely, such folly can't survive long, and GIS will force a change. Surely, a new heyday will come in this new age of geographic capability and awareness; globalization; and rising interest in spatial, integrative and place-based science.

(This column originally appeared in the January 2004 issue of GeoWorld.*)*

The Geographic Flaw That Toppled Communism

By Jerry Dobson

More than twenty years ago, when communism held uncompromising reign over eastern Germany, a village teacher discovered a secret more ominous than those imagined in the spy movies of the day. Browsing through a local library, Herr Rolf-Peter Bartz chanced upon a book that contained forbidden information. It had been a part of the university curriculum as recently as the early 1950s, then quietly dropped. There was no official campaign, nothing that drew attention, so it is not surprising that a copy remained in a provincial library. Bartz quickly recognized that the principles espoused in the book revealed a vital flaw of communism and that the events had taken place only a few kilometers from the village where he has lived and taught. The heart of the book is a simple geographic model—perhaps the most famous quantitative model among geographer's in the West—a model so simple, so elegant, and so fundamental that it is taught in virtually every introductory economic geography course, taken for granted among geographers and regional economists, and ignored by practically everyone else.

The Model

In the 1830s Johann Heinrich Von Thunen changed the way we look at our world (Von Thunen's *Isolated State*, An English Translation of *Der Isolierte Staat*, New York: Pergamon Press, 1966). Generally each farmer, from his personal vantage point, reckons the economic forces (market prices versus production, labor, and transport costs) and decides what his farm alone will produce. In current GIS terminology, we might think of this as a point-to-area relationship. Thunen, a farmer himself, took a holistic view. He wanted to know which crops should be grown at every point surrounding each market. The full array of points creates areal patterns so his view consisted of area-to-area relationships. These patterns are much too complex for analysis without GIS technology, so Thunen imagined an isolated city on a uniform plane. This gave him only one market, again a point-to-area relationship but from the opposite point of view.

For most farmers the reckoning consists of a diverse mixture of calculation, intuition, and mimicry; but Thunen wanted hard data. He kept a meticulous record of every cost and revenue. From these data he devised an economic rent formula that some have called the first quantitative model in geography and economics. I won't repeat the equation here. Many of you already know it by heart, and those who don't will benefit from reading the more complete accounts that can be found in economic geography texts, such as James O. Wheeler and Peter O. Muller. *Economic Geography*. New York: John Wiley & Sons, Inc., 1986 or Truman A. Hartshorn and John W. Alexander. *Economic Geography*. Englewood Cliffs: Prentice Hall, 1988. A GIS version (IDRISI, DOS) has been written by Rustin F. Dodson (VT/GIS: The von Thunen GIS Package, National Center for Geographic Information and Analysis, Technical Paper 9127, August 1991). Briefly, economic rent is analogous to the profit that can be gained by producing a particular commodity on a particular parcel of land. Transport cost and distance to market are key variables determining the spatial distribution of economic rent for each commodity. The result is a series of different crops (land covers) in rings surrounding the market. The ultimate goal is efficiency of spatial organization.

Spatial Efficiency

The principles involved are fundamental to every location decision in agriculture and can be extended to most other land uses as well, yet Thunen's model is often ignored outside the fields of geography and economics. Knowing what the model is designed to do, we may ask, "How can city planners draw land use maps without running the model? How can farmers and agricultural policy makers decide which crops to grow? How can engineers site agricultural processing and transport facilities?" The answer is simple. In a free market system prices force everyone (farmer, engineer, policy maker) to comply with the principles of the model. Farmers and entrepreneurs who do not meet the minimum thresholds of economic rent, either through conscious decision, intuition, or dumb luck simply fail, and their enterprises are replaced. Planners and policy makers who do not comply meet swift and resolute resistance.

Spatial inefficiency was a crucial factor in the failure of communist agriculture because there were no real prices to guide decision making. Bartz recognized this failure, but he hoped the deficiency could be overcome through education and informed decision making. With missionary zeal, he set out to inform fellow citizens that spatial efficiency is essential to survival. He brought the message home by quoting Thunen's adage that a portion of each crop is eaten by the wheels.

Spatial efficiency was anathema in a system where labor was paramount to all other factors. Cautiously and patiently, Bartz delivered his message for more than two decades. He began by riding his bicycle to the estate of Tellow where Thunen had developed the empirical model. He took his students to the estate and organized a restoration effort that continues today. He established the Thunen Museum. People came to see antique farm equipment and to let children ride ponies, but no one left without a dose of Thunen. Behind Thunen's barn, Bartz constructed concrete circles and used them to illustrate Thunen's rings. Simultaneously, Bartz assiduously collected data so that the spatial economy of his era could eventually be compared to that of the earlier free market system.

The disparity between the model estate of the 1830s and the communist collective of the 1980s would have been comical had it not been so detrimental to human welfare. The very land where Thunen had emphasized efficient transport became part of an even larger collective, and all farm equipment was stored at a central location and driven to distant fields each day. Workers were transported from central quarters to distant fields, to lunch, and then to other distant fields. Thunen had emphasized the efficient distribution of fertilizer from the dairy to nearby fields of intensive crops. In contrast, the collective land was divided administratively so that cattle and crops were institutionally separate. Fertilizer produced in the dairy often went unused while other fertilizers were imported from distant locations. Extrapolating these circumstances to the rest of the communist world, it is easy to understand why the grocery shelves have been empty for so long.

A Challenge to GIS Technology

A driving motivation of the Information Revolution is the implicit assumption that people can make better decisions if they have better information and faster, more comprehensive analytical tools. Here from Tellow is a speculative test: If the leaders of the collective and the communist nation had possessed modern GIS technology with uncommonly good spatial databases covering market prices and production, labor, and transport costs, could they have reached a satisfactory level of spatial organization? My guess is that the system could have been made somewhat more efficient, but the effort would still have been thwarted by centralized decision making. GIS technology is not even close to attaining the computational capacity to simulate the economic and physical forces operating at each and every parcel of land in a large region. GIS analysts' cannot possibly possess all the working knowledge of the farmer, the entrepreneur, the engineer, and the policy maker. The message is a warning against technocratic arrogance for we face the same constraints in the West. The best use of GIS technology is through decentralization so that improved information and tools are available at all decision making levels from bottom to top.

The Thunen Museum Today

I visited Tellow in April 1991 at the urging of Robert W. Peplies of East Tennessee State University who visited a year earlier before the Berlin Wall came down. Both of us were impressed by the story of this man, Bartz, and by the commitment that he, his family, and the people of the district feel for the continuance of the estate. It is a living museum and laboratory: the fields are still being cultivated and the data are still being collected. With pride, my hosts presented a meal of vegetables grown on the farm and wild boar and venison from nearby. The Parish Priest showed me Thunen's grave and his seat in the church. He spoke knowledgeably of Thunen's social philosophy and the role he had played in the community.

Currently Bartz is leading a critical effort to gather the records of neighboring collectives before they are lost in the chaos of abandonment. Simultaneously he leads the effort to protect the estate in the new order. There is still a danger that some new leader, unaware of Tellow's importance and beset with other crises, will destroy it with the stroke of a pen.

If Bartz's story dealt with secret weapons or political intrigue, it could be the making of an epic movie about commitment and heroism. But a story about the geographic flaw that toppled communism? Too esoteric! That would never sell!

(This column originally appeared in the May 1992 issue of GIS World.*)*

An Evil King Used Geography to Enslave the Congo

By Jerry Dobson

One of the great crimes against humanity was the enslavement of the Congo by King Leopold II of Belgium. And, if there is any crime for which geographers should issue a collective "mea culpa," it is this.

A Geographer of Sorts

From his youth, Leopold was obsessed with a personal quest to own and operate a colony—not for his country, but literally for himself and certainly not for the people of his chosen land. To find the right place, he studied world geography so diligently as a young man that, were he not an evil king, we might consider him an accomplished geographer.

Even as a child, he focused on geography to the detriment of other school subjects. In his late 20s, shortly before ascending the throne, Leopold spent weeks in Seville's General Archive of the Indies, studying records of the Spanish conquest of the Americas. He then spent months traveling the world and conducting fieldwork that might have earned him a master's degree or possibly a doctorate in geography if he had been properly enrolled. His thesis might have been titled, "An Economic Geography of Global Factors Affecting the Profitability of Colony Ownership."

None of this, of course, can be blamed on the discipline. There came a time, however, when Leopold enlisted an entire world community of nobles, scholars, clergy, humanitarians and politicians in his evil scheme, and he exploited geographers in particular. To put it bluntly: he duped us, and we fell for it wholeheartedly.

Most of us have come of age in a time when government is run by lawyers and economists; industry is run by MBAs and engineers; and science is dominated by physicists, chemists and other specialized disciplines. Geographers are without power. "GIScientists" influence, but don't control, information policy, and they have no voice in broader policy circles. We've come to think of the situation as normal, but it hasn't always been so. In Leopold's day, geography as a discipline and geographers as individuals were powerful to an extent that we hardly can imagine today.

Evil Disguised as Good

In September 1876, the King of the Belgians convened an international geographical conference in Brussels. Its stated purpose was "to consider ... the scientific exploration of the unknown parts of Africa and the best method of penetrating into the interior of that continent with a view to its civilization and the abolition of the slave trade" (Wright, 1952). "His majesty invited the leading African explorers, several eminent geographers and the heads of the geographical societies of Great Britain, France, Germany, Italy, Russia and the United States" (Wright, 1952).

Geographers, explorers and humanitarians alike were greeted like visiting royalty, housed in the Royal Palace, and charged with planning and promoting the king's campaign to open trade routes, establish permanent bases, force peace among natives and abolish slavery in the Congo. The participants endorsed Leopold's plan and created the International African Association with Leopold as its chair. No one questioned his intentions, and Leopold was hailed as a great humanitarian.

A Clever Deception

Today we question the concept of "civilizing" indigenous societies, but Europeans in those days considered their own brand of civilization far superior to any alternative. It was their gift to the world, whose self-evident benefits justified all means of imposing it. However, "civilization" alone wouldn't have attracted the support that Leopold sought. The issue that galvanized the world in his favor was slavery. Europeans and Americans, having recently abolished slavery in their own territories, were eager to attack the Arab slave trade still under way in East Africa. For the next three decades, Leopold deviously exploited their anti-slavery fervor to justify and conceal unspeakable acts of cruelty directed by him against the people of the Congo.

In reality, the new philanthropic organization functioned for little more than a year, but Leopold used its name and flag for almost 10 years, deviously alternating with the International Association of the Congo—owned outright by him. Politicians became so thoroughly confused that, when the United States in 1884 became the first nation to recognize Leopold's sovereignty over the Congo, its decree named both organizations synonymously in a single sentence. As soon as his sovereignty was guaranteed, Leopold dropped all mention of the philanthropic organization and exclusively used the name of the corporation that he controlled. Ultimately, he named his private domain the Congo Free State, one of the greatest misnomers in geography.

In today's information age, it's difficult to imagine how isolated the African interior was during Leopold's reign. Firsthand accounts of atrocities emerged as early as 1890, but events in the Congo were as unverifiable as "canals" on Mars, which were accepted as scientific fact by far more believers at that time. This veil of geographic isolation, aided by Leopold's control over entry up the Congo River, gave him and his concessionaires a free hand to do what they liked with native populations. Whole districts were depopulated, and as many as 10 million people died, first in the frenzy to extract ivory and later for rubber.

The litany of horror is too vast to describe in this brief column. To gain a sense of it, read Adam Hochschild's (1998) *King Leopold's Ghost* or Joseph Conrad's

(1902) *Heart of Darkness*. A worldwide humanitarian campaign finally ended the "rubber terror" in 1908, when Leopold sold his private domain to Belgium at an enormous profit.

The American Role

Reading Hochschild's book, I was curious to see what role American geographers might have played. Leopold's invitation to the 1876 conference could only have gone to the American Geographical Society (AGS), because no other geographical association existed in America at the time. Indeed, the invitation did arrive, but the AGS Council took no action and no one made the trip to Brussels.

However, Judge Charles P. Daly, president of the society from 1864 to 1899, initiated and presided over the American branch of the philanthropic International African Association. In that capacity, he joined Henry Shelton Sanford, former U.S. minister to Belgium, in seeking U.S. recognition of Leopold's claim (Wright, 1952). Every other action in Judge Daly's long tenure at the AGS helm shows him to have been good and wise, but he was misled along with scores of other world leaders by one of the greatest deceivers history has ever known.

Power and Responsibility

I fully expect GIS soon to confer upon geographers, GIScientists and GIS practitioners greater power than geography has known since World War II. Information really is power, and it can be used for good or evil. What happened in the Congo and the geographers' unintended contribution to it are stark reminders of the responsibility that comes with such power.

(This column originally appeared in the July 2002 issue of GeoWorld.*)*

Can GIS and Geopolitics Prevent War?

By Nigel Waters

According to recent articles on the Web and in *The New Yorker*, the Bush Administration is conducting clandestine reconnaissance operations in Iran with a view toward pre-emptive strikes on nuclear facilities. Geopolitics seems to be back in the news.

In my own department, in an attempt to purge unnecessary courses and concentrate on our core strength of GIScience, we almost cancelled our only course on political geography. Saving it may be a prescient decision, because geopolitics soon could become the most important course a geography department can offer—especially if it's combined with the tools of GIScience.

Geopolitics Then

The term "geopolitics" has its roots in the work of Halford Mackinder, who advocated the geostrategic advantages of land power over sea power. Specifically, he suggested that whoever dominated the Heartland of the Eurasian landmass would dominate world politics (Mackinder, 1962).

According to Mackinder, the balance of sea power and land power has been history's dominant theme as empires have risen and fallen. Balance has ensured peace; predominance of one over the other has led to war. Knowing how to subvert the balance has led to the rise of empires from Egypt to 19th-century Germany.

One of the early battles between sea and land power was between the Greeks (and their colonies by the sea) and Xerxes of Persia. To render Greek sea power impotent, Xerxes marched in from the north after crossing the Dardanelles on a bridge of boats in 480 B.C. The Greeks fought to heroic defeat at Thermopylae. Thebes capitulated, and Athens was burned before Xerxes was himself defeated in the naval battle at Salamis.

Alexander the Great exercised his revanche in 334 B.C. when he reversed Xerxes' route. Inexplicably, his army's crossing of the Hellsespont went unmolested. After crossing the sea, Alexander first destroyed Persian sea power in the eastern Mediterranean. Then, having rendered this impotent, he attacked the land-based armies of the Persian Empire.

Mackinder recounts much of the world's history in terms of the struggle between sea- and land-based powers. Particularly revealing is his discussion of the great figures of European politics during the 19th century: Napoleon, Talleyrand,

Metternich and Bismarck.

Mackinder had grudging respect for Bismarck, and he was fascinated by Bismarck's genius for diplomacy. This was never more clearly seen than in the Austro-Prussian annexation of the duchies of Schleswig and Holstein in 1864.

The great European powers of Austria, France and Russia had been brought onside by collusion, bribery and mutual antipathy toward Poland, respectively. Denmark's sole ally, the sea power England, was unable to deliver on Prime Minister Henry John Temple Palmerston's promise of aid, because this was a land-based war. Thus, a chain of events was set in place that led directly to the global wars of the 20th century for domination of the Heartland—the so-called "Great Game."

Geopolitics Now

Despite my own department's tepid interest in geopolitics, interest elsewhere is thriving. Brian Blouet (1987) wrote a definitive biography of Mackinder's work (and has another promised for 2005). In his biography, Blouet notes Mackinder's linking of military interests and geography as well as his membership in a "war game" club.

Other writers have described post World War II criticisms of Mackinder's theories. According to W.H. Parker (1982), the critics are numerous. But Mackinder also has his defenders, who declare with him that "the most fundamental of all elements of national power is sheer location on the globe."

Parker notes that Richard Nixon referred to the United States as an "island" country and a "sea power" and the Soviet Union as a "land power" that was "situated in the center of the Eurasian Heartland."

Perhaps this is a case of academic jealousy over a theory in which the generals and politicians have given credence to in varying degrees. It's certainly true that many learned from Mackinder's observations.

Karl Haushofer took Mackinder's ideas that were ignored in the West and used them to advise Stalin, Hitler and Japanese militarists. Mackinder noted the vulnerability to attack of sea bases from land-side armies. Haushofer pointed this out to Japanese military strategists planning the subsequently victorious 1942 land-based attack on Singapore, where the guns all faced seaward.

Mackinder gave us much more than a theory. He was a great believer in academic disciplines that led to synthesis, and he abhorred academic specialization. He did much to establish the academic discipline of geography in British universities, including founding the School of Geography at Oxford University in 1899.

He emphasized, like Albert Einstein, the need for imagination in scholarly research. Mackinder suggested that this would be a necessity until geographers had their own "space machine" for making these imaginations a reality—and what's a GIS if not the "space machine" of Mackinder's dreams?

Space Machines

To be effective, any GIS needs spatial and attribute data. Mackinder noted the importance of geography to the German public of the late 19th century: "every educated German is a geographer in a sense that is true of very few Englishmen or Americans." Such knowledge of geography eventually was linked to the German school of Geopolitik and was used to justify German expansionist philosophies in the early 20th century.

A better use for space machines is to imbue them with real attribute and spatial data and then simulate the results of expansionist efforts. The results typically aren't what the simulators would want, so caution rules the day.

An exercise of this sort was carried out recently by James Fallows for the *Atlantic Monthly* (*www.theatlantic.com/doc/prem/200412/fallows*). The intelligence dilemma to be investigated with Sam Gardiner, a war games expert, was the following: "What should the United States do about the expectation that Iran would have 'deliverable' nuclear weapons in 2008?"

Considered options included a punitive raid against Revolutionary Guard units, a pre-emptive strike on nuclear facilities and, third, a regime-change operation involving the forcible change of the government in Teheran to a pro-Western administration. This final option would involve the first two as a prelude. The war games came with an impressive series of PowerPoint maps that, according to the participants, did much to increase the exercise's "cool factor."

One can only speculate on how much better the war games would have been if they were supported by a full range of GIS analytical and visualization tools. Regardless of the "cool factor," Gardiner concluded with two sentences: "You have no military solution for the issues of Iran. And you have to make diplomacy work."

However, an Iran that knows this is unlikely to cooperate, and perhaps that's why we hear from *The New Yorker* that covert operations inside Iran are continuing.

Prophylaxis to War

Others can play war games, too. Russia must be concerned about recent events in the Balkans, Iraq and Iran (Alexander's original Heartland) as well as now in the Ukraine. The Heartland is in danger. Even more disturbing to Vladimir Putin must be the fact that the aforementioned war games had U.S. air strikes being launched from Azerbaijan and Afghanistan.

Should we be surprised to hear that Russia and China are again planning joint military exercises? If the Heartland is so badly compromised, is it any wonder that it seeks to join with China to restore balance by becoming a sea power? Darker scenarios are being considered, including those described in *Liberating Taiwan*, a book recently published by the Chinese Military Publishing House (*www.jamestown.org/news_details.php?news_id=74*).

If, however, every government is using war games and space machines, perhaps we can move toward a world in balance that Mackinder envisaged. Perhaps the space machines can be used as a prophylaxis to war.

(This column originally appeared in the March 2005 issue of GeoWorld.*)*

How Canada Gave GIS to the World and Then Lost Almost Everything

By Nigel Waters

There it sits in the middle of the computer output from my last multiple-choice test: "Question 7" with a big asterisk beside it. An asterisk means that less than 25% of the students got the question correct, which would be expected if they just guessed the answer and didn't have a clue which of the four responses was right. In fact, only 8% of the students had the correct answer; they weren't even close to 25%.

What went wrong? I'd never seen such a low-scoring question. I quickly checked to see if I had the answer wrong—I didn't.

The question was as follows:

7. The first GIS is generally recognized as having been developed in:

a. Canada

b. the United States

c. the United Kingdom

d. Japan

Answer "a" is, of course, correct. Canada, under the guidance of Roger Tomlinson, developed the first GIS for the Canadian Federal Government in the 1960s. It was known as the Canada Geographic Information System (Tomlinson, 1984). My students either lacked a little patriotism, or, given the current state of affairs in Canadian GIS education and research, perhaps they simply couldn't believe that Canada had been first.

They also may believe that Canada no longer is a leader in GIS, and that would be correct. But if Canada was so far ahead in the 1960s, what happened? Why did Canadian GIS lose its direction? There were three major opportunities that Canada missed.

Canada Never Created a GIS Center

In the late 1980s, the National Science Foundation (NSF) in the United States and the Chorley Royal Commission in the United Kingdom investigated the development of national centers of excellence to boost GIS research and dissemination. In the United States, NSF funded the National Center for Geographic Information and Analysis (NCGIA), which has been an enormous success in sponsoring research and educational initiatives. There's little doubt that the NCGIA helped cement the U.S. position as world leader in GIS technology. The center's distribution of low-cost teaching materials for GIS instruction in universities and, more recently, on the World Wide Web (*www.ncgia.ucsb.edu/education/ed.html*) has made NCGIA's influence pervasive.

In the United Kingdom, recommendations from the Chorley Commission helped establish eight Regional Research Laboratories and two associated centers by the U.K. Economic and Social Research Council.

Canada never got its center. Instead, we did what Canadians have been doing for years, we borrowed and adapted research and educational initiatives pioneered in the United States and the United Kingdom. In my own teaching, I used the NCGIA Core Curriculum and Understanding GIS (ESRI, 1990), a book that originally had been developed in part at Birkbeck College, University of London.

It's true that Canadians now have a relatively well-funded ($30 million of funding between 1999 and 2002) GEOIDE center at Laval University. Although it brings together, according to promotional material, "25 universities, 26 companies, and 16 government agencies and departments," it seems to be too little, too late and, for many GIS researchers, somewhat irrelevant.

Canada Charges Cost-Recovery Prices for Data

Canada, and admittedly the United Kingdom and many other countries (Corey, 1998; Rhind, 1998; Yeh, 1998), always has been an aggressive proponent of a cost-recovery approach for disseminating GIS and related data. Although the debate has raged long and loud on this topic (Onsrud, 1998), it seems that the U.S. approach has stimulated a hugely successful GIS industry, while the Canadian approach has allowed organizations such as Statistics Canada to bleed universities dry of their ever-diminishing operating funds.

Again, Canada recently made steps in the right direction by establishing Statistics Canada's "Data Liberation Initiative." However, it's still expensive to participate in this program. Moreover, the restrictions on data use and the draconian penalties for misuse place a chill on research conducted within Canadian universities. After a decade of buying Canadian data less expensively from U.S. companies, rather than their own government organizations, it's hard for Canadian researchers to break old habits, including using free U.S. data to teach Canadian students about a technology Canada invented.

What's In a Name?

Canada hasn't convinced the rest of the world to use "geomatics" as a "catch-all" phrase for spatial technologies. Canadians like the word, because it's so similar in French (géomatique) and English—an important concern in a country that's officially bilingual.

The GIS industry, like no other enterprise, has spawned a proliferation of names and acronyms (see "What's In a Name?" *GeoWorld*, November 1998, page 40). More than a score of the names can be found at the University of Colorado Geography Department's Web site at *www.colorado.edu/geography/gcraft/gloss/glossary.html*.

True, they're all synonyms for GIS, and geomatics aims at something broader, but now we have the term "GIScience." In French, this must be "IGScience," but the difference is rather minimal, as there aren't even any accents to worry about.

But using the same or similar words within Canada also is a matter for debate. When I suggested to a close Francophone (French-speaking) friend from Montreal that it was good to use a word such as geomatics, which is similar in both languages, he replied that the English should get their own word. By contrast, in France, at the Academie Française, and in Quebec, Canada, at the Office de la Langue Française, they have made it their mission to purge French of English words such as "weekend." For the French, the problem is especially bad with respect to recent, technical terminology (such as GIS). To halt the English onslaught on their language, the Quebec organization publishes the Grand Dictionnaire Terminologique (*www.lgdt.cedrom-sni.com*) and Le Français au Bureau, which help Francophones realize that what just appeared on their computer wasn't E-mail—it was "courriel."

When I suggested an alternative to "GIS" at a meeting of the Executive of the Canadian Association of Geographers more than a decade ago, I still remember Roger Tomlinson turning to me and saying, "Nigel, why don't you call it what it is, a Geographic Information System."

How Can Canada Get Back Its Groove?

Canada has made some progress with the adoption of "geomatics." A quick Internet search shows that it's popular in Australia and the United Kingdom, and linguistic variations may be found in France, of course, as well as Greece, Sweden, Austria, Germany (where it's "Geoinformatik") and Spain, among others. Even some U.S. government Web sites, such as Alaska's Bureau of Land Management, now use the term. However, I still prefer GIScience.

Concerning human resources, we may expect the "brain drain" of Canadian GIS researchers to the United States to continue and perhaps accelerate. This is in part due to the weak Canadian dollar, government cuts to academia and more generous American research grants. One ray of hope is the recently announced Canadian Federal Government program that seeks to establish 1,200 new professorial chairs at Canadian universities. If some of these new chairs are dedicated to GIS teaching and research, and if they're woven together into a virtual "center" of GIS excellence (perhaps following technical suggestions in Ramroop and Pascoe, 1999), then maybe the next time I ask my students which country invented GIS, more than 8% will know the correct answer.

(This column originally appeared in the January 2000 issue of GeoWorld.*)*

Hire an Anthropologist to Find the Future of GIS

By Nigel Waters

Developments in GIS have been closely related to developments in computing and computer science. The domination of mainframe computers throughout the 1970s meant that GISs were cumbersome and awkward to use. GIS was a "nonstarter" for all but the largest industries and institutions until the 1980s ushered in the era of the "microcomputer" and a transformation in GIS software's ease of use (for a brief history of GIS, visit the Web at *www.geog.ubc.ca/courses/klink/gis.notes/ncgia/u23.html#UNIT23*).

However, GISs remain relatively difficult to use, and true software fluency and understanding is acquired only after extensive exposure. This situation may change, thanks to the Massachusetts Institute of Technology (MIT) Oxygen system.

What's Oxygen?

Oxygen is a package of technologies being developed at MIT to make computing more "user friendly" and, therefore, more useful in our daily lives. It's surprising that we still need these types of advances after 50 years of commercial computing technology. But Michael Dertouzos, director of MIT's Laboratory for Computer Science for the last 25 years, notes that computers definitely can be made easier to use (Dertouzos, 1999; also see *www.sciam.com/1999/0899issue/0899dertouzos.html*).

Dertouzos is no slouch with computers; neither is Tim Berners-Lee (you may recall that he invented the Web) or Ronald Rivest, co-inventor of the Rivest-Shamir-Adelman algorithm for public key cryptography. However, not one of them—seat mates on a recent plane ride to Taiwan—could get Dertouzos' laptop operating system to interface properly with a plug-in card. Even the newly hired faculty whiz kid, who "graciously" offered to help the old boys, failed. Success was achieved only after three hours of trial and error by Dertouzos. I'm sure most GIS analysts can relate to this sort of nonsense, yearning for the day when they can just tell their GIS what to do. Oxygen should help.

Dertouzos describes Oxygen as a series of eight complementary technologies:

1. Cell-phone sized, hand-held computers with visual display, infrared detectors and a camera.

2. In-situ "wall" computers.
3. Dynamic intranets that allow a selected "region" of users to interact in collaborative fashion.
4. Broad-domain speech understanding.
5. Personal and worldwide knowledge access.
6. Collaborative technology that links personal and/or job-related actions taken in a temporal fashion (sounds suspiciously like a note-taking system my father devised 30 years ago).
7. Automation of routine tasks.
8. Customizing the system to individual needs.

What's Wrong with This Picture?

Each piece of Oxygen merely is an extension of existing technology—as are most innovations (remember the Sony Walkman?). As Dertouzos notes, technologies' real power lies not in individual novelty, but in integration into a useful, seamless whole. The Walkman was a runaway success, because Sony realized that technology should fit easily into the way people want to live their lives. The devil, however, often is in the details.

But how can one ensure that an enormous investment in hardware design actually would help people do what they want with greater ease and less frustration? Hiring an anthropologist may be a good start.

Recently in Canada, journalist Diane Francis, editor and columnist for *The Financial Post* (Francis, 1998), and David Johnson, minister of education and training for the provincial government of Ontario (Johnson, 1998), trashed social-science students as being a waste of taxpayer money, urging greater disparity in government grants to engineers and computer-science students. During 1998 in Ontario, each social-science student brought in $3,500 (CDN) and each engineering or computer-science student generated $7,000 (CDN) of provincial government funding (Guerette, 1998).

Some researchers argue that social-science students are irrelevant in a high-tech world. But in reality, the opposite is true. Social-science students are more essential than computer-science students. To design a system properly, even a computer system, you have to observe people in their work environment. No one is better trained to observe, record and understand individual behavior and cultural practices, at work or at home, than anthropologists.

Recent articles (Koerner, 1998; Church, 1999; Hafner, 1999; Weise, 1999) state that although anthropologists rarely are seen aiding in computer system design, they're being welcomed by major corporations throughout North America. Of the 2,000-plus anthropologists working outside academia in the United States, more than 40 percent work as consultants to the business world (Koerner, 1998), often for corporate giants such as Nestle, Motorola, Intel, Hewlett-Packard, Apple and AT&T Labs.

Susan Squires, president of the National Association for the Practice of Anthropology *www.ameranthassn.org/napa.htm,* notes how anthropology has developed and evolved methods such as participant observation and other ethnographic procedures (Miles and Huberman, 1994; Denzin and Lincoln, 1994) that are specifically designed to understand how people live and work in a variety of cultures (Weise, 1999).

The famed Xerox Palo Alto Research Center, for example, hired anthropology graduate Lucy Suchman in 1979 to observe how people copy documents. Suchman's work led to the single, green copy button—the "just do it" button. This was a blow against rampant featurism and a victory for simplicity (Norman, 1998). Have you ever tried to find the green print button on your computer keyboard? Don't hold your breath. In many networked systems, printing a document is one of the more non-intuitive things you can do. Suchman now is somewhat of a legend in the field of applied anthropology and was a featured speaker at a recent Community of Anthropologists of Science, Technology and Computing conference *http://symptom.mit.edu/castac.*

A Parting Shot

I believe Francis is misguided when she attacks the social sciences in general and anthropologists in particular by stating, "Degrees not in demand—whether it's history or anthropology or law—should be more expensive [for the student]."

So go out and prove her wrong. Hire an anthropologist and double your GIS business, design a better GIS or both. The Institute for the Future in Menlo Park has a whole staff of ethnographers studying technology's effects (Weise, 1999). Perhaps Francis works for the Institute of the Past.

(This column originally appeared in the November 1999 issue of GIS World.*)*

Section 2.
References

Blouet, B. 1987. *Halford Mackinder: A Biography*. Texas A&M University, College Station, Texas.

Bowden, M.J. 1970. "John Kirtland Wright, 1891-1969," *Annals of the Association of American Geographers*, Vol. 60, No. 2, pp. 394-403.

Church, E. 1999. "Firms Seek Guidance from Anthropology," *Globe and Mail*, July 26, 1999, M1.

Corey, M. 1998. "Digital Data, Copyright, Commercialization and Related Issues—A Canadian Perspective." in Taylor, D.R.F. (ed.), *Policy Issues in Modern Cartography*, pp.29-46. Pergamon, Oxford, England

Denzin, N., Lincoln, Y.S. (eds.). 1994. *Handbook of Qualitative Research*. Sage Publications, Thousand Oaks, Calif.

Dertouzos, M. 1999. "The Future of Computing," *Scientific American*, Vol. 281, No. 2, pp. 52-55.

Dobson, J.E. 1979. "A Regional Screening Procedure for Land Use Suitability Analysis," *Geographical Review*, Vol. 69, No. 2, pp. 224-234.

Dobson, J.E. 2001. "ORNL Created a Powerful Legacy for Geographic Research," *GeoWorld,* Vol. 14, No. 9, p. 26.

ESRI Inc. 1990. *Understanding GIS: The ARC/INFO Method*. ESRI Press., Redlands, Calif.

Fogel, R.W. 1964. *Railroads and American Economic Growth: Essays in Economic History*. The Johns Hopkins University Press, Baltimore, Md.

Foresman, T. (ed.) 1998. *The History of GIS: Perspectives from the Pioneers*. Prentice Hall, Upper Saddle River, N.J.

Francis, D. 1998. "Our High-Tech Brain Cramp: Canada's Universities are not Producing Enough Engineers," *National Post*, Oct. 31, 1998, D3.

Guerette, J. 1998. "Nortel Stresses Cooperation Between Industry, Education and Government to Tackle the Need for More High-Tech Education."

Hafner, K. 1999. "Coming of Age in Palo Alto: Anthropologists Find a Niche Studying Consumers for Companies in Silicon Valley," *New York Times*, available online at: *www.nytimes.com/library/tech/99/06/circuits/articles/10anth.html.*

Hammond, A. (ed.) 1981. *King Richard III (The Arden Edition of the Works of William Shakespeare)*. Methuen, New York.

Hammond, P.W. 1993. *The Reputation of Richard III. Richard III: A Medieval Kingship*, Gillingham, J. (ed.). St. Martin's Press, New York.

Hochschild, A. 1998. *King Leopold's Ghost*, Houghton Mifflin Co., New York.

Huntington, S.P. 1998. *The Clash of Civilizations and the Remaking of World Order*. Simon and Schuster, New York.

James, P.E., Martin, G.J. 1970. *The Association of American Geographers: The First 75 Years*, AAG Publications, Washington, D.C.

Johnson, D.J. 1998. "Ontario Government Supports High Tech," *National Post*, Nov. 9, 1998, C6.

Jones, D.C. 1976. *The Court-Martial of George Armstrong Custer*. Charles Scribner's Sons, New York.

Koerner, B.I. 1998. "Into the Wild Unknown of Workplace Culture: Anthropologists Revitalize Their Discipline," *U.S. News*, Aug. 10, 1998.

Limp, W.F., Cothren, J. "It's a 3-D World!" *GeoWorld*, Vol. 16, No. 1, p. 30.

Mackinder, H.J. 1962. *Democratic Ideals and Reality*. Norton, New York.

Miles, M.B., Huberman, A.M. 1994. *Qualitative Data Analysis*. Sage Publications, Thousand Oaks, Calif.

Milton, G. 1996. *The Riddle and the Knight: In Search of Sir John Mandeville*. Allison and Busby, London.

Norman, D.A. 1998. *The Invisible Computer: Why Good Products Can Fail, the Personal Computer Is So Complex and Information Appliances Are the Solution*. MIT Press, Cambridge, Mass.

Onsrud, H.J. 1998. "Tragedy of the Information Commons." In Taylor, D.R.F. (ed.), *Policy Issues in Modern Cartography*, pp.141-158. Pergamon, Oxford, England.

Parker, W.H. 1982. *Mackinder: Geography as an Aid to Statecraft*. Clarendon Press, Oxford, England.

Ramroop, S., Pascoe, R. 1999. "Implementation Architecture for a National Data Center." In Goodchild, M.F. et al. (eds.), *Interoperating Geographic Information Systems*, pp. 64-74. Springer-Verlag, Berlin.

Rhind, D. 1998. "National Mapping as a Business-like Enterprise." In D.R.F. Taylor (ed.), *Policy Issues in Modern Cartography*, pp. 1-18. Pergamon, Oxford, England.

Smith, N. 2003. *American Empire*, University of California Press, Los Angeles.

Taylor, D.R.F. (ed.). 1998. *Policy Issues in Modern Cartography*. Pergamon, Oxford, England.

Tomlinson, R.F. 1984. "Geographic Information Systems - A New Frontier." *The Operational Geographer,* No. 5, pp. 31-36.

Waters, N. 1998. "Geographic Information Systems," *Encyclopedia of Library and Information Science*, Vol. 63, pp. 98-125.

Weise, E. 1999. "Anthropologists Adapt Technology to World's Cultures," *USA Today*.

Wright, J.K. 1952. *Geography in the Making: The American Geographical Society, 1851-1951*, American Geographical Society, New York.

Yeh, G. 1998. "Availability and Pricing of Georeferenced Data in Asia Pacific." In D.R.F. Taylor (ed.), *Policy Issues in Modern Cartography*, pp. 47-69. Pergamon, Oxford, England.

Section 3.
GIS, Spatial Analysis, and the Art of Mapping

Introduction
By Daniel Sui

Stripped of all the bells and whistles of geospatial technologies, there should be no disagreement that spatial analysis and mapping in real-world coordinate systems remain to be the two primary functions that distinguish GIS from other software packages. To conduct spatial analysis and visualize the results in various map forms are the two main tasks of any GIS projects, after the digital databases are created and properly geo-referenced. In fact, GIS technology was originally developed as a technology to automate the tedious manual process of mapping in the 1960s and 1970s, and the integration of GIS with spatial analysis and modeling dominated academic research in the 1980s and throughout the 1990s. Within less than 50 years, it is indeed a remarkable technological accomplishment for the powerful geospatial technologies that are widely used all over the world today. Geospatial technologies are poised to march forward for an even bright future with the two powerful legs of analysis and mapping.

The columns in this section address issues related to analysis and mapping. Dobson and Waters argued for the need of more sophisticated analysis and modeling capabilities to be embedded in GIS. Using examples of developing a new Neandertal theory and habitat conservation, Dobson and Waters further illustrated the importance of spatial analysis and modeling in scientific research and for addressing environmental concerns. To Dobson, scientific integration can better be achieved through the synthesis of information in a GIS. Columns by Crampton and Sui challenged some of the existing cartographic/mapping conventions in GIS. In addition to 3-D visualization, other alternatives such as cartograms, indigineous mapping etc. are discussed.

Fully in line with the artistic tradition in the cartography, contributors in this section all argue that neither analysis nor mapping is necessarily a dreary, boring process. Instead, one has to develop a keen sense of artistry and a deep appreciation of beauty in order to do both analysis and mapping correctly. Waters', "Ten Most Beautiful Formulaes in GIS," hopefully will inspire readers to seek more beauty in GIS analysis whereas Sui's attempt to link GIS with Beethoven and Picasso will motivate our readers to conduct GIS analysis and mapping in the tradition of a humanistic GIScience. Indeed, spatial is special, special is often crucial, and crucial is always cool.

Although selective, we hope these essays will stimulate your interest to ponder more deeply about the analytical and mapping dimensions of GIS.

A New Spatial Statistic Is Needed to Estimate "Probability-of-Fit"

By Jerry Dobson

What's the probability that two lines, independently drawn, will happen to have the same shape? Existing statistical methods are inadequate for estimating such a probability. Techniques are available to measure the "goodness-of-fit," but not the likelihood that a given fit will occur by accident—let's call it "probability-of-fit."

This is a rather surprising omission in spatial statistics, considering the many uses that could be made of such a measure.

Georegistration of Images and Maps

A probability-of-fit algorithm could help, for example, in georegistering aerial photographs or satellite images without ground control points. Sets of curved lines that appear on separate images or maps could be tested to see which ones are likely to represent the same line.

Yang Cheng and Michael Hodgson (1996) developed a georegistration algorithm that matches ground control lines as well as points, but the approach requires candidate lines to be identified manually. An analyst must select several lines in each axis so the image can be adjusted in both directions. A probability-of-fit statistic would automatically select candidate lines that could be matched, and the images then could be georeferenced through an algorithm. Cheng and I discussed this possibility and decided that such an algorithm would be desirable, but it may be too computationally demanding for routine use in image processing.

A renowned geographer once told me that he needed a probability-of-fit algorithm to compare ancient descriptions of shorelines with current hydrographic charts. A classic case would be a treasure map crudely drawn long ago. What's the chance that an ancient, poorly described feature matches a suspected coastline today?

Continental Drift

My primary interest focuses on what constitutes evidence in the ongoing debate over continental drift. When Alfred Wegener (1912) publicly proposed continental drift based on the matching coastlines of South America and Africa, he was ridiculed for the rest of his life and 30 years after his death. The 1960s discovery of sea-floor spreading finally brought respect to his theory, but not to his evidence, logic or methods.

About 10 years ago, I spoke with a prominent geologist who still claimed that Wegener was wrong—not that the theory itself was wrong, but rather that Wegener's use of coastlines and other geographic patterns didn't constitute a contribution to the theoretical foundations of plate tectonics. His view was that Wegener wasn't doing science at all, merely voicing a lucky guess based on inadmissible evidence.

I was prepared for a similar reaction when I publicly proposed two other fits of South America and Africa, a new fit of South America and North America and new fits of Australia and eastern North America (Dobson, 1992, 1996). The response hasn't been ridicule, but rather a mighty "ho hum."

To date, not a single geologist or geophysicist has cited either of my papers. Why? Because coastline fits are considered to be trivial accidents without scientific merit. That's been the standard counterclaim since Wegener published his views, and presumably since geographer Abraham Ortellius first proposed continental drift in 1596 (Romm, 1994)).

Indeed, prejudice against that line of evidence and logic is so great that other supporting evidence is ignored regardless of how convincing it may be in its own right. Wegener presented many other geographic patterns (glaciation, polar wandering, etc.) besides the coastlines themselves. For my Australia/North America fit, I presented tectonic, paleontologic and lithologic evidence that matches in time, space and process far better than the evidence for any of the accepted fits.

In addition to the traditional fit of South America and Africa, I found four major configurations and numerous intermediate positions. The number and quality of fits presents two alternative interpretations:

1. Meaningless continental fits are "a dime a dozen," and Ortellius' and Wegener's original fit was purely accidental.

2. Accidental fits are rare, and these new configurations may indicate additional continental migrations that were previously unrecognized.

Thus, probability-of-fit is key to interpretation. It would be thoroughly unscientific to reject them as accidents without a proper measure, but that's been standard practice for decades—and possibly for centuries.

Measuring Probability-of-Fit

In the case of continental drift, it's reasonable to consider each match as a set of two lines, one from each continent. These could be coastlines, continental shelves, other bathymetric or elevation contours, geologic formations or any other recognizable lines. They need not even be at the same elevation if there's suspicion that vertical rise or fall may have occurred on either continent.

I believe the odds against a perfect, accidental match are long. Obviously, an algorithm must consider goodness-of-fit, extent of departure from a straight line and the spatial resolution at which lines are measured. Perhaps there should be more than one goodness-of-fit: one to measure general trends and another to measure detailed correlations. Matching convoluted lines would seem less likely than matching straight lines.

I've considered various ways to construct the algorithm, but none seem satisfying. Is it a random walk with 1,000 steps left, right or forward? I've tried it that way, and the negative odds reach astounding numbers. Even when divided by the total number of world coastlines of similar length and adjusted for the fact that the matches aren't truly perfect, the numbers remain huge.

Perhaps it's best to imagine a single origin and destination with a multiplicity of alternative pathways to connect them. In this case, what's the probability that two unrelated physical processes will yield the same pathway? But again, how does one reflect the similarity and the curvature of the two chosen pathways?

Or perhaps it's best to imagine a potential surface defined by two axes, one representing the length of matching lines and the other representing their maximum departure from a straight line. Within that surface, the two lines can occupy any space, constrained only by the fact that adjacent points on any given line can't lie farther apart than one spatial unit at any given resolution.

I offer this challenge to spatial statisticians everywhere: Invent this algorithm, and the world will beat a path to your door. But what's the probability that any two of those paths will be exactly the same without some underlying causal connection?

(This column originally appeared in the July 2000 issue of GeoWorld.*)*

Geographic Analysis Yields New Neandertal Theory

By Jerry Dobson

The first Neandertal (see "Author's Note") was discovered in Germany in 1856, and they've caused trouble ever since. For the first 30 years, several key scientists doubted Neandertals were a separate, ancient species. When similar skeletons of unquestioned antiquity were discovered in the late 1880s, the weight of evidence tilted in favor of recognizing Neandertals as modern humans' close relatives. For most of the 20th century, anthropologists have debated the closeness of that relationship.

The "replacement" theory holds that all Neandertals died out and were replaced by anatomically modern humans. An opposing view, "continuity," holds that some intermarriage or other sexual contact occurred, thus ensuring that certain Neandertal traits continue in humans today.

At the heart of this debate, the following recurring questions arise: How could Neandertals be so similar to modern humans, and yet be so different? How could Neandertals have occupied Europe for longer than modern humans are known to exist (roughly 200,000 vs. 100,000 years), and yet disappear in a brief 10,000 years? And how could modern humans replace Neandertals so quickly and completely with no sign of struggle? Collectively, these and other questions comprise the Neandertal enigma. Even the most respected Neandertal experts openly plead for new ideas that may help solve these mysteries.

Shaving Neandertal with Ockham's Razor

How similar are Neandertals to modern humans? In 1939, Carleton Coon drew his impression of a Neandertal neatly shaved and wearing modern clothing. William Straus Jr. and A.J.E. Cave later claimed that Neandertals, thus trimmed and dressed, might pass unnoticed in a New York subway. However, morphological differences occur in virtually every bone, and these are fundamental differences affecting length, shape, thickness and weight. If these differences are due to genetics alone, sizable numbers of genes must have mutated during the divergence of the two species from a common ancestor. The causal mechanism and multiple factors that influenced so many mutations would have been highly complex.

William of Ockham, an English scholar who lived from 1285 to 1349, liked to keep things simple. He proposed a rule, known as Ockham's razor, which is interpreted to mean that the simplest answer is usually the best. Ockham's razor would favor a single factor and mechanism to explain the multiple differences between Neandertals and modern humans.

In *Geographical Review* (Dobson 1998), I offer a simple explanation that accommodates what's known about Neandertals and their successors. The single factor that accounts for the differences appears to be iodine, and the mechanism of change may have been a single genetic modification which improved the efficiency of modern thyroid glands to extract iodine.

A comparison of Neandertal anatomy with that of modern humans who suffer from iodine deficiency disorder (IDD) reveals far too many similarities to be coincidental. Distinctive Neandertal traits—overall body proportions, heavy brows, heavy musculature, dental development and wear, and propensities for degenerative joint diseases—are identical to those of modern humans suffering from cretinism, the most severe form of IDD. Late Pleistocene human evolution, consequently, may be explained by several alternative interpretations that involve iodine pathology and/or biological adaptation. Speciation may have resulted from geographic isolation of interior vs. coastal populations.

Venus Figurines

The Europeans who displaced Neandertals are the Cro-Magnons. Their culture included artistic innovations such as cave painting, sculpture and ceramics. Conspicuous among their artifacts are tiny statues and bas reliefs of squatty females with huge breasts and bellies as well as conspicuously exposed genitalia. These "Venus" figurines exhibit distinctive traits associated with cretinism among modern humans. Their overall body form and some detailed characteristics, such as the shape of fingers and nipples, are consistent with the deformities of mature female cretins.

Cro-Magnons evidently revered the cretinous form, but what does that imply about their relationship with Neandertals? Perhaps there's no connection. Or perhaps the first Cro-Magnons sighted the last Neandertals and passed that knowledge to future generations through the artistic expression of Venus figurines. Or perhaps Cro-Magnons descended directly from Neandertals, and later generations considered the inevitable cretins of their own age to be reincarnations of their ancestors.

A Question of Geography

Whenever I present these findings, the most common response is, "How did you get into this line of research?" My answer illustrates the value of geographic thought and analysis. I didn't start out to study iodine or Neandertals or Cro-Magnon figurines. I started with fundamentally geographic questions about projected sea-level rises. Global warming has led some people to worry about future rises of a few feet. Looking into this issue, I learned that sea level has already risen 125 meters since the ice ages. The vertical rise and horizontal extent are astounding. This fact caused me to ask, "If the coast of 18,000 to 5,000 years ago is missing, what does that imply about the archeological record?" Agriculture, for example, first appeared about 10,000 years ago in three different forms in three separate places. Was there a common precursor down on the old coast?

This line of speculation led my brother, Jeff Dobson, and me to ask, "Is there some geographic factor that might distinguish coastal populations from interior populations?" It was Jeff who first thought of iodine, which is essential for modern humans and is almost entirely a coastal resource derived from saltwater fish, shellfish and seaweed. Even sea salt doesn't contain enough iodine unless purposely fortified. Today, about 30% of the world's population is at risk of IDDs. Indeed, 750 million people suffer from goiter, 43 million suffer IDD-related brain damage and mental retardation, and 5.7 million are afflicted by cretinism.

My brother learned that the most iodine-deficient regions on Earth today are "mountainous and formerly glaciated," which led me to ask, "What about Neandertals? They lived in mountainous Europe during the ice ages. Did they need iodine? If so, how did they get it?" I compared Neandertal and cretin morphology and ultimately concluded that Neandertals were iodine-deficient.

GIS Is Needed

I couldn't have performed my research without the vast body of evidence tediously amassed by anthropologists, archaeologists and paleontologists. For 150 years they labored, literally in the trenches, to understand Neandertals and Cro-Magnons, and geographers such as Karl Butzer of the University of Texas at Austin contributed mightily. For an even longer period of time, physicians and medical investigators have struggled to understand goiter and cretinism, aided by geographers such as Harold Foster of the University of Victoria, British Columbia, Canada. Presently, however, there exists no definitive, comprehensive, multiscale geographic analysis of Neandertals or Cro-Magnons. Perhaps cruelest of all, there isn't a single map of iodine's global distribution at scales suitable for human health studies. Scientists must rely on mapping fellow humans who suffer from iodine's absence. These fields desperately need the integration, spatial logic and multiscale perspective that geography and GIS can bring.

If any *GeoWorld* readers are interested in developing the world's first GIS devoted to Neandertals, prehistoric art, iodine or modern IDDs, I'll help and advise any way I can.

Author's Note: In its original German language, the spelling was "Neanderthal," but the proper pronunciation always was "täl." "Thal," meaning "valley," was one of many words spelled "th" but pronounced "t." The German language reform of 1900 changed many such spellings to conform with pronunciation. Today, most German, French and American scholars prefer "Neandertal," while many British scholars prefer "Neanderthal." The general public in the United Kingdom and America typically employ "th" in spelling and pronunciation.

(This column originally appeared in the November 1998 issue of GeoWorld.*)*

Scientific Integration Means GIS

By Jerry Dobson

The greatest need in science today is for improved integration among disciplines. Calls for it grow louder each year, especially at the highest levels of science policy, but no one at the top seems to recognize what integration really means. Synonyms abound—"synthesis," "multidisciplinary," "interdisciplinary," etc.—but rarely is the term "geography" mentioned in those circles.

The Essential Integrator

Take any set of physical processes (e.g., geologic, biologic, chemical, atmospheric, etc.) and any set of cultural processes (e.g., social, economic, political, etc.) and ask yourself what accounts for the interactions that occur among them. The obvious answer is place. No matter how different the processes are, they will interact if and only if they occur in the same place. Indeed, they can't avoid interacting when they occur in the same place.

Change the land, and you change the culture. Change the culture, and you change the land. The connections are obvious at local levels everywhere on Earth, and they apply at all scales, but science and society are just beginning to understand how they work at the global level. Anyone who tries to integrate physical and cultural processes, regardless of his or her disciplinary background, will have to reckon with the same spatial factors that GIS is designed to handle. It can't be done without geometry, topology and attributes; points, lines and areas; networks and flows; and connections and adjacencies.

Science communication and decision support also require interpreting language into mathematics, then mathematics into visualization enhanced by data and summarized in words. For centuries, traditional, analog cartography helped substantially, but it didn't satisfy this need, especially when communicating with scholars, decision makers and others not trained in geography or cartography. In recent decades, GIS has revolutionized our ability to integrate diverse data. Now it's also beginning to integrate models.

GIS serves as a common denominator among diverse interests working within any given region. Often it forces individuals of diverse disciplines, fields and responsibilities to talk with one another. That's true from higher education and research to the workaday world of local government and business. Daratech, a leading market-survey firm, recently characterized current trends in the global market for GIS as "new disciplines, industries, markets and applications." Such widespread societal movement toward GIS raises hope for a unifying tool that bridges scientists, analysts, decision makers, operations managers and the public.

STEP Linkage

From a geographic perspective, it's useful to think pragmatically in terms of linking several different types of models across space, time, entity and process (STEP). Process models are the type aspatially employed by most specialized disciplines. One example would be a biologic model used to simulate tree growth. Typically, geographic factors are expressed as state variables (e.g., temperature and moisture) or as input variables (e.g., nutrients) that may be assumed constant, specified interactively by model operators or acquired from external models.

Spatial relationships usually aren't explicit, except that there's a growing body of cell-based (i.e., raster data structure), 2-D models, especially in ecology. Almost without exception, however, their ability to handle spatial relationships is limited to direct hand-offs between any given cell, for example, and one of its immediate four or eight neighbors. Even the simplest GIS greatly exceed such limited representations of spatial relationships.

Transport models are employed by climatologists to simulate air movement, hydrologists to simulate water flow and transportation geographers to simulate highway or rail flow. For linear networks such as railroads and highways, spatial relationships usually are best represented as nodes (points) and lines in vector data structures, preferably with specified connections, directions and interchanges. For area-wide flows such as air movement, spatial relationships often are cell-based (raster) with feature identities, connections, directions and interchanges expressed as attributes of cells. Data exchange is limited to each cell and its four or eight neighbors, and longer transports are handled as a domino-like cascade from cell to cell across an entire landscape or region.

Transportation models also may incorporate some elements of process, but they tend to emphasize processes closely tied to the movement itself. A hydrologic model, for example, may deliver water of a specified temperature and pH to a given cell, but to grow a tree with that water in a particular cell would require far greater spatial- and process-modeling capacity than any transport model currently possesses.

Much Work Is Needed

For all its power, GIS still is a fledgling technology. There remains an overriding need for software and hardware systems that can link STEP models. Initial efforts to develop a STEP linkage system indicate that it will be computationally intensive, requiring simultaneous processing of many models covering many topics and sub-areas (polygons or cells) within each region. The integration of such disparate models is a technical challenge of historic proportions, but it pales in comparison to the difficulty of integrating the paradigms, institutions and cultures of diverse disciplines and professions.

Pardon me if I grow weary of calls for integration from high-minded, well-intentioned scientists who have risen through the ranks of specialists and don't understand anything I just said.

(This column originally appeared in the May 2002 issue of GeoWorld.*)*

What Are the 10 Most Beautiful Formulae in GIS?

By Nigel Waters

Earlier this year, on the basis of its engaging title, I purchased a copy of the book, *The Most Beautiful Mathematical Formulas* (Salem et al., 1992). The book is a delight, but it does not live up to its intriguing title. It struck me that it would be challenging to come up with the most beautiful formulae in GIS.

The Top 10 List

To be included, these formulae have to have some of the following criteria: they must be widely used; they should be fundamental building blocks of many key GIS algorithms; they should impart powerful ideas; and they should represent and relate to other key formulae.

1. The Oldest of them All

Perhaps the most beautiful of all is the oldest. It is the formula for distance between two points and is based on Pythagoras' Theorem from the sixth century B.C.

$$D = \left[|X_1 - X_2|^2 + |Y_1 - Y_2|^2 \right]^{1/2}$$

The vertical bars indicate the absolute value and the 1/2 indicates the square root. If the 'twos' in all three exponents are replaced by R then R is known as the Minkowski metric. When R=1 we have the so-called city block or Manhattan distance. This formula has been used ingeniously in location-allocation modeling and is an important building block in GIS.

2. The Intersection of Two Lines

$$X_i = -(a_1 - a_2)/(b_1 - b_2)$$

$$Y_i = a_1 + b_1 X_i$$

This formula providing the X and Y coordinates of the point of intersection is another beauty from the field of coordinate geometry, or as GIS techies say, "COGO." It is the cornerstone of many point-in-polygon and polygon overlay routines, and where would GIS be without them? The b terms represent the slope of the first and second lines, and the a terms represent the Y intercepts.

3. The Gravity Model

$$I_{ij} = k(P_i)(P_j)/d_{ij}^{\,2}$$

This was used to predict traffic flows in the earliest transportation models from the 1950s. Quite simply, it states that the interaction, I, between two places i and j is equal to some coefficient, k, multiplied by their population product divided by the distance between them squared. It can be extended and manipulated to produce potential models important in marketing and popularized in GIS packages such as SPANS, now owned by PCI, Inc., Richmond Hill, Ontario, Canada.

4. The Entropy Maximizing Model

$$T_{ij} = A_i O_i B_j D_j / e^{bc_{ij}}$$

This was the basis of the entropy maximizing models developed in the late 1960s and provided statistical respectability for the trip distribution model, because it was possible to show that the model predicted the most likely distribution of trips. Oi and Dj are the number of workers in the origin zone and number of jobs in the destination zone, Ai and Bj are weights, e is the irrational number 2.718..., b measures the friction of distance and c is the cost of travel.

5. Projection Formulae

$$X = R(l - l_0)$$
$$Y = R \ln \tan(p/4 + f/2)$$

Fifth on the list are formulae for converting spherical coordinates to Cartesian X, Y values. In the above equations l and f are longitude and latitude, respectively, and X and Y are the standard Cartesian coordinates. R is the radius of the sphere and l_0 is the central meridian with all angles measured in radians. Finally, p is the mathematical number pi. This elegant formula (from Snyder, 1987) for the oldest of all standard map projections has been used to produce maps of the planetary surfaces of most of Earth's nearest neighbors.

6. Least Squares Models

The formula for a straight line in Cartesian space is also the formula for the common regression line, and this regression equation can be extended easily into two independent dimensions to produce an interpolated trend across space:

$$Z = a + b_1X_1 + b_2Y_2$$

Thus, the equation above is a first order polynomial with two independent variables. If the dependent variable is elevation and the two independent variables are eastings and northings, then the regression equation provides a linear, interpolated surface portraying the trend in elevation. From this we can move easily to higher order surfaces that allow us to model more complex relief. Also, we can add another independent variable and plunge into the third dimension—a powerful formula indeed. With just one independent variable we can model how our dependent variable, for example economic rent, changes over distance, and we have the classic bid rent line—Von Thunen's model.

7. Optimization Models

$$Maximize\ Z = aX_1 + bX_2\ subject\ to\ constraints$$

Here we have the basis for all the programming models in their various guises, including linear, integer and dynamic programming and the so-called transportation models. All of these are crucial if we wish to use our GIS as a spatial decision support system.

8. The Rank Size Rule

$$P_i = P_1 / l$$

This is a wicked one with all sorts of statistical implications and applications. A hint of the treasure trove awaiting the investigation of this formula is given by Simon (1991, pp. 372-375). The formula says big things are less common than little things, and it describes log-normal distributions, which are common in the data we store within a GIS. In this version, P_i is the population of the town of rank L and P_1 is the population of the largest town.

9. The Poisson Distribution

$$p(n) = e^{-d} d^n / n!$$

Here d represents the density of the process and p (n) is the probability of observing n points in a given interval. This formula has been used to evaluate the distribution of samples of points to see if our GIS has provided adequate coverage when the data are spaced irregularly. It can be used to generate data for traffic simulation models where vehicles arrive independently of each other and in many other situations.

10. Classification and the F Statistic

$$F = Within\ class\ variance / Between\ class\ variance$$

So simple, so elegant and yet it is the basis of most of our attempts to classify the data in our GIS, and classification is fundamental to our use of GIS in resource management using remotely sensed data.

What's Your Opinion?

These formulae were selected for their elegance, simplicity, generality and their wide applicability and power. They represent a whole host of approaches and analytical procedures that have been fruitful in the construction of GIS and in the subsequent analysis conducted using GIS.

(This column originally appeared in the November 1994 issue of GIS World.*)*

Can GIS Save Banff from the "Invigorating Scent of Dollar Bills?"

By Nigel Waters

GIS use for environmental modeling and management of parklands has been an unparalleled success by any measure. Events that showcased this research include the Integrating GIS and Environmental Modeling conferences, which were held in Boulder, Colo., in 1991, and Breckenridge, Colo., in 1993. These events resulted in two popular texts: *Environmental Modeling with GIS* (Goodchild, Parks and Steyaert, 1993) and *GIS and Environmental Modeling: Progress and Research Issues* (Goodchild and Steyaert, 1996). A third conference was held in Santa Fe, N.M., in 1996, and the conference proceedings may be obtained on CD-ROM directly from the National Center for Geographic Information and Analysis at *www.ncgia.ucsb.edu/conf/SANTA_FE_CD-ROM/santa_fe.html*.

A fourth Integrating GIS and Environmental Modeling meeting was held in Banff, Alberta, Canada, from Sept. 2-8, 2000. Full details of the event may be found on the Web at *www.Colorado.EDU/research/cires/banff/*. Banff, for many reasons, is an ideal location for such an international conference on GIS and the environment.

Banff under Siege

The town of Banff is situated in the heart of Banff National Park, which, when it was designated Canada's first national park in 1885, was only the third national park in the world. The United States' Yellowstone National Park was first, and Banff occupies a similar spot in Canada's national psyche. Banff, together with its sister Mountain Parks, Jasper, Kootenay and Yoho, and three nearby Provincial Parks, form the Canadian Rocky Mountain Parks World Heritage Site (Krakauer, 1995).

However, all is not well in Canada's mountain paradise. Popular articles, such as Jon Krakauer's *National Geographic* expose, "Rocky Times for Banff," discuss the division among those who wish to protect the environment and those who wish to exploit it for the "invigorating scent of dollar bills" (to use Sid Marty's phrase, which is quoted in Krakauer's article). Krakauer, now a best-selling author, interviewed internationally renowned wildlife biologist and wolf specialist Paul Paquet, who explained that the best habitat for wolves and their prey is low elevation, riparian valley land. This area, unfortunately, also is the preferred habitat for human settlements, including Banff.

Remarkably, Banff National Park includes Canada's two major transportation corridors, the Trans-Canada Highway and the Canadian Pacific Railway (CPR). These routes snake their way through what are ideal wildlife transportation corridors (Alexander et al., 1996). The passageways' attraction for wildlife has been exacerbated by grain spillage from CPR railcars. Herbivores are attracted to the grain spills, and herbivores get hit by trains. Wolves are attracted to the herbivores, and wolves get hit by trains. It's all pretty simple.

Krakauer also presents opposing opinions. There's the unusual view of Banff's former senior wildlife warden, Rick Kunelius, that development makes for an improved ecosystem: "Development has actually increased habitat for wild ungulates by creating more forage: the golf course, the recreation grounds, everybody's yard."

A Banff Springs Hotel clerk is quoted as saying: "If those tree huggers manage to prevent the twinning of the highway, the blood of innocent crash victims is going to be on their hands." Apparently, the clerk had never heard of realistic speed limits.

Debate in the Local Media

Debate about the future of Banff and the park's role has resulted in a running battle in the local media. One concern is about opening the Lake Louise ski area gondola during the summer months. Parks Canada has bravely recommended against this, because the area during the summer is a prime feeding ground for grizzlies. Steve Herrero, who now is completing the five-year Eastern Slopes Grizzly Bear Project (*www.canadianrockies.net/grizzly*) and is arguably the continent's most renowned grizzly bear expert (Herrero, 1985), has applauded this decision (*Calgary Herald*, Sept. 29, 1999, A19).

But about two weeks later, Kunelius wrote to the same newspaper, criticizing Herrero for noting that of the 107 grizzlies killed or removed between 1971 and 1996, all were within "500 meters of a road or developed area, or 200 meters of a high-use trail." (Such statistics can easily be acquired through routine GIS buffering operations.) Kunelius makes the quite ludicrous statement that one old grizzly apparently dying of tuberculosis near a road doesn't make roads dangerous, and claiming they are hazardous is "akin to claiming that more people die in bed than anywhere else, therefore beds are a serious mortality factor" (*Calgary Herald*, Oct. 15, 1999, A29). With attitudes like that, it's a wonder there are any animals left in the park.

The Debate in Fiction

An intriguing addition to the "Banff Wars" is Graeme Pole's *Healy Park* novel (1998). Pole is a respected, award-winning author of nonfiction books that relate to the Canadian Rockies, and he recently published his first novel about the crisis in Banff National Park. Despite all the typical disclaimers that only a few characters are inspired by living persons, "insiders" will tell you they know "who this character is" and "which combination of individuals that character is."

All of Banff's complex stew of government overlords, park administrators, wildlife wardens, commercial adventurers, wildlife researchers and environmental activists are in the book, along with the media and those who just want to hike in peace.

There's even a railroad that seems to drop a lot of grain along its tracks. There's a "wolf man" and a "bear man" (I wonder who they could be?) and even a "wolf pack" of dedicated researchers with ties to the University of Calgary. There's also a public relations "spin doctor" whose motto is "no bad news" and who wants to review all wildlife research before it reaches the media.

At one point in the "novel," Pole argues that twinned, four-lane highways are more dangerous than the old two-lane roads. He's particularly critical of the mitigation philosophy, referring to it as the "Humpty syndrome," which involves allowing the destruction of the ecosystem followed by elaborate attempts to repair and restore it. Classic examples of this are the culverts and underpasses that Pole argues are ineffective. Because of their feebleness, Parks Canada had to install, at a cost of $4 million, "two wildlife overpasses, whose efficacy was virtually untested elsewhere."

Healy Park is pretty black and white. There's no question who the good guys are—the wildlife researchers and those who want to hike, leaving nothing but their footprints—and who the bad guys are—the park administrators who wish to bring more tourists into the back country to increase the park's commercial potential. But even Pole's somewhat elitist model is nonsustainable, because, as Krakauer notes, on a warm summer's day there may be so many hikers that a path may quickly become "a boggy, 20-foot-wide cattle trail." The city of Calgary is only an hour away, and, even without the tourists, there soon will be a million people sitting on Banff National Park's doorstep.

Debate in Academia

The issues mentioned in this column have resulted in two conference series that confront these intractable problems. First, there are the Roads, Rail and Environment Conferences that specifically address the effects of transportation infrastructure on wildlife and other aspects of the environment. Second, there are the International Conferences on Wildlife Ecology and Transportation, which recently released its third set of proceedings (Evink et al., 1999). A growing number of papers presented at these conferences now are "couching their research" within a GIS framework.

It's also hoped that when the GIS and Environmental Modeling researchers roll into Banff in September, they'll be able to use their expertise in GIS research to shed more light and less heat on these topics that have proved so contentious.

(This column originally appeared in the March 2000 issue of GeoWorld.*)*

Beethoven, Picasso and GIS: Is Spatial Really Special?

By Daniel Sui

Despite the fact that geospatial technology has become a multi-billion dollar business during the last decade and was recently named one of three defining technologies (along with bio and nano) for the 21st century, it still doesn't have a mantra that resonates and inspires the public.

If I could nominate one, I think "spatial is special" should be the catchphrase for the geospatial industry.

I like this line because it goes beyond the technical and scientific meaning it carries. Perhaps more importantly, I believe that this ambiguous, poetic line can stir the public imagination along multiple dimensions. It's fascinating to notice that spatial is indeed special in many interesting ways.

Music to the Ears

To be special, a technology must carry unique meanings beyond common levels. Some of my colleagues in academia have gone the extra mile to show how spatial is special.

I'm intrigued by Keith Clarke's (2001) analogies between raster data structures and the musical notes of Mozart or between vector data and Beethoven's music.

Like the music of Mozart, rasters are meticulous, repetitive and highly structured. In contrast, vectors are bold and abrupt, often leaping from one step to another swiftly, just like the music of Beethoven.

Maps can be perceived as music frozen in space, and music envisioned as maps flowing in time. Humans have an almost universal love for music, so perhaps there's no better way to enthuse students than presenting GIS as music to their ears.

Paintings to the Eyes

Maps and mapping have been closely linked to visual arts throughout the history of cartography. Throughout the ages, cartographers have been concerned about how mapping could become a better art for presenting spatial data and how art could be used as an inspiration to make better maps.

But the emergence of GIS, and especially GIScience, isolated GIS from the spirit of visual arts. But science and art shouldn't be separated. Similar to the musical connections between GIS and Beethoven or Mozart, I found works by two 20th century artists—M.C. Escher and Pablo Picasso—linked to spatial representations in many fundamental ways.

Escher's drawing technique, for example, is based primarily on spatial tessellation—recursively subdividing space into various shapes and then using these shapes to reconstruct reality.

In contrast, Picasso, in his representative cubist works, fragmented the subjects he tried to paint into points, lines and polygons, and then he reconstructed them into sets of interlocking patterns. Thus, cubists built up a more object-based view of the world (a vector-based representation in GIS terms).

Humanistic GIScience

So far, the development of GIScience has been squarely grounded in traditional science and technology. In my opinion, the next phase of GIScience research and education should be broadened and involve fields in arts and humanities to make spatial truly special.

Humanistic GIScience attempts to integrate human knowledge with key issues related to spatial data representation, analysis and visualization. Instead of emphasizing accuracy by trying to minimize or eliminate uncertainty, humanistic GIScience adds the human subjective and imaginative dimensions of experience to facilitate the processing and understanding of spatial data.

The goal of a humanistic GIScience is to build dialogs with a variety of different scholarly traditions to develop a better understanding of reality's complexities.

As the quest for new means of analysis and modeling via computers has been increasingly intertwined with the persistent search of the deepest meanings of such activities, we should expect to witness a revitalization of aesthetic and humanistic traditions within GIS and cartography.

Artists' renditions of the world—real and imagined—in novels, poems, paintings, movies, music and songs can be rich sources of inspiration for GIScience researchers and students exploring alternative conceptualizations of space, place, time, environment, region and scale.

With the infusion of aesthetic traditions into GIS, we can anticipate that humanistic GIScience will flourish in the near future. In addition to the common representations of space framed by Euclidean geometry, humanistic GIScience is attempting to find novel ways to handle the textures of place as articulated in the humanistic tradition as well as the structures of space.

(This column originally appeared in the November 2005 issue of GeoWorld.*)*

Examining the Art of Mapping

By Jeremy Crampton

Is cartography art or science? This is an age-old question that's undergoing re-examination by a series of cartographically minded artists. It seems that the swing toward science is being complemented by an emerging swing toward art.

An example is Kanarinka, a Boston-based artist whose social activities hover almost equally between the virtual and the physical (Kanarinka is an online name that has become her regular name). At a recent talk to North American Cartographic Information Society members in Portland, Maine, Kanarinka, co-director of the nonprofit collective iKatun (*www.ikatun.com*), described three interesting geo-artistic projects that challenge our way of thinking about time and space. The projects could be called "psychogeographies."

According to geographer David Pinder, these psychogeographies are about more than the banality of not finding one's way, but are a deliberate disorientation and perhaps reorientation toward the alienating city. They owe much to the Situationists of the 1950s and 1960s.

Going Dutch

In one such project, participants were invited to explore New York City using a map of Copenhagen. The artists claimed that "the maps of central New York and Copenhagen are surprisingly identical ... if we look at the two cities as innumerable movable layers that can be imposed above or below other layers ... [it is] perfect for an investigation in the field of placement and re(dis)placement." (Perhaps not quite what the GIS manuals have in mind when they write about map overlays.)

Another project focuses on the very small. The Analysis of Infinitely Small Things (*www.ikatun.com/k/infinitelysmallthings*) is named after a real work on geometry. By drawing attention to the artifacts and objects that are normally below our radar screens, the artists think we might be able to refocus on the city or at least see it through fresh eyes.

The project includes a guidebook and a research kit. The guidebook is based on the geometry text but in a total reinterpretation. For example, the text "find any reflected ray" becomes an instruction to "capture" light being reflected in a sample packet. Another example was "find a table," which became all sorts of found objects sitting on tables.

Fleeting Funerals

Another project was called "funerals for a moment," which commemorates the trivial, unsung moments we experience every day, such as hearing someone sneeze or a man yell out. The idea is to hold a "funeral" for these fleeting, irrecoverable moments at the exact space they occurred.

A group of mourners visits the spot and carries out the wishes of the person who experienced the moment. This might involve a moment of reflection or yelling. We mourn the major things in life such as someone dying, but aren't these major things just comprised of many smaller things? Isn't it a question of scale?

Noticing the Unnoticeable

Many of these projects revisit themes from the humanistic geography of the 1970s. I'm reminded of Denis Wood's short essay on scale in the book *Humanistic Geography*. He discussed going to the store and buying a carpet. While you're in the store, the carpet seems large, perhaps too large for your room. Then when you get it home and throw it down, it seems to become all too small—until you drop your contact lens on it, and it becomes a dauntingly vast landscape as you search on hands and knees.

This kind of art is fun, and that's important, especially if you are teaching mapping and GIS to students. It gets people to establish a new relationship with the spaces and geographies around them, and perhaps notice the almost unnoticeable.

But art also can be serious—an intervention, a call to action or a protest. Picasso's Guernica is just one example of an artist finding political expression through art. The recent upsurge of "cartographic art" is indicative of some emergent need to deal with the spaces of our lives.

These mapping projects may force us to change our way of thinking about maps. Instead of a map being a document that says "this is how the world is," we see that maps are (also? instead?) documents that say "this is how the world is not ... or how it could be."

(This column originally appeared in the January 2005 issue of GeoWorld.*)*

Are Choropleth Maps Good for Geography?

By Jeremy Crampton

One of the most common forms of mapping data today is the "choropleth" map, in which each area (e.g., state or county) is shaded according to the data. Where there's more of something, the area will be shaded more darkly and vice versa. Technically, a choropleth is based on a stepped statistical surface, and an isarithmic map is based on a smoothed surface.

After you start looking around, choropleths are everywhere. Students love choropleths, and they're also standard map options in GIS software. I'm sure many readers have made choropleth maps. But have you ever wondered where they came from and when they were invented? In fact, they had a kind of "double invention," with the person who actually coined the name warning people from using them.

First in France

The first known choropleth map was constructed in 1826 in France, which was a hotbed of thematic mapping (Robinson 1982). After the choropleth was introduced by Baron Charles Dupin, there was a "mini craze" to depict social problems with choropleths. Dupin's first map was of education rates around France. He was an early education reformist and not a cartographer. He linked the health of the nation and the education of the population—what's good for the nation is therefore necessarily good for the population.

In 1829, another set of choropleths was published, showing crimes and educational levels, by Adriano Balbi and André Michel Guerry, Italian and French statisticians who had an interest in "moral statistics" or what we would now call social problems (e.g., crime, education, birth rates, suicide, etc.). These men were political economists, not cartographers. Crime and education maps reveal where problems are in your country and therefore where to deploy resources. In today's terms, they're maps that speak to the security of a territory and its resources.

Named in New York

Neither Dupin nor Balbi and Guerry used the term "choropleth" to describe their inventions. The term had to wait another hundred years—until 1938—before it was invented. The man who invented the term was J.K. Wright, then with the American Geographical Society (AGS) in New York City (Wright 1938).

The AGS was (and is) an important society that provided expertise to the government during the two World Wars. The superbly named "Chief Territorial Specialist" was Isaiah Bowman, AGS director. Bowman hired Wright as librarian in 1920, but Wright had tremendous cartographic interests. In 1932, he edited Paullin's famous historical atlas of the United States, which later would form the model of the national atlas published by the U.S. Geological Survey in 1970.

Wright coined the word "choropleth" in a document called "Notes on Statistical Mapping" by AGS and the Population Association of America, which had just been founded. Wright chose to focus on "mapping population phenomena and other data on the basis of statistical information." This was difficult and innovative. No one had a systematic understanding of thematic or statistical maps or had yet invented the rules for using them. Today, cartography textbooks have whole chapters on choropleth maps, but it wasn't until 1938 that the first English-language cartography textbook was published (Raisz 1938).

Wright's Warning

Wright (Wright 1938) says, "the term choropleth, which expresses the idea 'quantity in area,' is tentatively proposed. A choropleth is an areal symbol ..." Choropleth literally means "area" and "fullness, quantity" (as in plethora). Wright presumably based his word by analogy to the much older word "isopleth," which is a line of equal quantity. Wright also classified cartographic symbols for the first time into our now familiar "point, line, polygon" division in 1938.

Wright tried to warn us off choropleth maps—especially where they show large areas, because they hide a lot of internal variation, and he dismissed them as inferior "cartograms,"by which he meant graphics that did not rise to the level of proper maps. Wright favored another map type, the lesser known "dasymetric," which can capture data patterns that occur across and within enumeration units. Unlike choropleth maps, which are based on geopolitical boundaries (county and state boundaries for example) dasymetric map boundaries are formed by the data itself. This allows a much more accurate mapping of population for example. Although mapping and GIS software aren't yet capable of directly producing dasymetric maps in the same way as choropleth maps, we might take Wright's advice and question whether choropleth maps—despite their popularity—are really telling us the whole story.

(This column originally appeared in the January 2003 issue of GeoWorld.*)*

3-D Geovisualization Is Coming of Age

By Jerry Dobson

There's something alluring about 3-D geovisualization that goes beyond its inherent utility. But there are many questions. How good is it? How feasible is it with commercial products on the market today? What progress has been made? What breakthroughs are needed?

Gee Whiz and Brute Force

Traditionally, the appreciation of 3-D geovisualization has come on two distinctly different levels. Not long ago, the principal effect of 3-D presentations lay in their "gee-whiz" visual impact on the uninitiated. There's just something about seeing the world as you would see it in person.

Implementation was cumbersome and expensive, however, and 3-D analytical software functions were practically nonexistent. Consequently, no one could afford to do it for more than demo effect. Many efforts sold themselves to sponsors based on mind-blowing demos, only to fail in utter disappointment when sponsors inevitably recognized the limitations.

Inside the GIS community, a corps of knowledgeable insiders recognized that 3-D GIS could be powerful if it was made more analytical. The key to analysis lies in creating the same data structures—geometry, topology and attributes—for the third dimension that we now have in 2-D GIS. Previously, the only way to accomplish such functionality, in practice, was through brute-force solutions involving as many as six different types of software:

- Image processing
- Raster/vector GIS
- CAD
- Virtual reality
- Spatial models
- Animation programs

That, in turn, required so many conversions and transformations that only a "crack team" of GIS specialists could make it work. Brute force still is essential, but breakthroughs as recently as spring 2002 reduced the conversion and transformation burden to a manageable level.

Virtual Cities

Mike Batty and Andy Smith (2002) estimate that worldwide today about 60 cities with populations greater than 1 million have "serious" virtual 3-D models. Most, however, don't have "serious" analytical functionality. New York City has one of the best systems—it's functional, and it proved its value on Sept. 11, 2001, and thereafter.

In the fall semester of 2001, I told graduate students in my advanced GIS seminar what had happened in New York. I showed them a 3-D geovisualization of Washington, D.C., done by a consortium of organizations led by the National Capital Planning Commission. I also showed them a 3-D geovisualization of airborne biological agents blown over Middle Eastern cities done by Richard Durfee, Ed Tinnel, Phil Coleman and others at Oak Ridge National Laboratory. When I tentatively mentioned producing a similar 3-D cityscape for Lawrence, Kan., the students' enthusiasm was such that I made it the focus of my spring seminar in 2002.

The result was "Larry," a 3-D GIS model of Lawrence produced by six University of Kansas (KU) graduate students in a single semester. They selected the software, collected the data and rendered a 3-D geovisualization and fly-through. Matt Dunbar and Hugh Howard focused on software assembly and application. Gary Hecox focused on field data collection and database development and, with Matt's help, measured the buildings in a two-block swath flanking the downtown street from end to end. Josh Campbell processed and interpreted high-resolution airborne imagery.

To prove the model's analytical value, Hugh Howard built a day/night population model, and Willem Helms built a flood model. Ivan Welch served as project manager and kept account of the labor involved in each task. The city of Lawrence generously provided planimetric data. The Kansas Data Access and Support Center provided digital elevation models and digital orthophotos, and the Kansas Applied Remote Sensing Program provided high-resolution experimental imagery.

3-D Software

There are many different kinds of 3-D geovisualization, and it's important to distinguish which ones fit each need. In our case, we wanted to show a continuous land surface as well as place buildings on the surface. We weren't interested in viewing the cubic space (rooms) inside, but architecture classes may want to do that in the future.

We tried several approaches and finally settled on two packages. For real-time display and interaction, ESRI's ArcScene served our needs well. We appreciated its seamless integration of diverse datasets and user interface. For 3-D rendering, we started with 3-D Nature's World Construction Set and later moved to the company's Visual Nature Studio. We're hoping for true 3-D GIS in a single package, but we're skeptical of all claims at present.

Effort Levels

Looking at the 3-D geovisualizations of New York and Washington, people naturally assume that only large cities can afford such efforts. Larry proves it can be done for small cities as well, although I would be the first to admit that Larry's crew wasn't typical of small-town workers. Still, it's a sure sign of 3-D GIS progress that six people could capture a city in one semester.

To be precise, we estimate the total labor to have been a little more than 1,000 hours with approximately one-tenth going to conceptualization and half divided evenly between execution and visualization. Data collection took about 57 hours, mainly spent in surveying buildings and their occupants. GIS software manipulation took about 85 hours, and visualization software manipulation took about 125 hours.

Payoff

Unquestionably, Larry generated a level of excitement among university, municipal and state officials that we never could have sparked with 2-D GIS alone. Most pleasing of all, Larry is attracting collaboration across campus, much like having a "real-world" laboratory available for experiments tailored to each discipline. A professor of computer science plans to adopt Larry in a future seminar and overhaul the graphic presentation. A professor of architecture plans to adopt the system and overhaul its buildings. Ultimately, Larry promises to place geography in the integrative role across campus that we've always claimed as a discipline, crossing all physical and social sciences.

(This column originally appeared in the January 2003 issue of GeoWorld.*)*

A Perfect Distortion? Cartograms Deserve More Attention

By Daniel Sui

In 2004, Michael Gastner and Michael Newman, two young physicists at the University of Michigan, made the cover story of the Proceedings of the U.S. National Academy of Sciences (see Plate 4). It wasn't because of their work in particle physics, but for their development of a new technique for making maps. Their work also captured a headline in Science News.

To be precise, Gastner and Newman didn't develop a technique to make ordinary maps—they created a new method to produce cartograms using a diffusion technique borrowed from particle physics.

Mathematically, making a continuous-area cartogram isn't a trivial process. Most of the existing algorithms often end up with results that lack readability or produce overlapping regions.

The Gastner-Newman algorithm uses a diffusion equation in the Fourier domain borrowed from elementary physics, which allows variable resolution by tracking moving boundaries. Compared to most of the existing algorithms, the Gastner-Newman algorithm can be executed with speed and gives readable results even for complex maps. Their software code can be downloaded at *www.santafe.edu/~mgastner/*.

Although an interdisciplinary group of scholars has been working on innovative algorithms during the last four decades, I'm amused by the fact that physicists achieved this major breakthrough. To my surprise (and, to some extent, disappointment), many GIS users are unfamiliar with cartograms or at least haven't paid enough attention to them.

Why Cartograms?

Cartograms are maps in which the size of geographic areas such as countries or provinces is represented in proportion to other properties (e.g., population, GDP, army size, etc.). Cartographically speaking, a cartogram can be defined as a non-conventional map projection that consciously distorts either area or distance (or both) to reveal patterns not readily apparent on a more traditional base map.

Because densities are equal on cartograms, they have been frequently referred to as density equalization projection (DEMP) maps in public-health literature. Although cartograms also are sometimes known as value-by-area mapping, they don't have to be area-based.

Charles Joseph Minard's 1861 map of Napoleon's military march into Russia or Harry Beck's 1933 London underground map were two of the well-known earlier linear cartograms. Although the first cartogram can be traced back to 350 A.D. (*www.livius.org/pen-pg/peutinger/map.html*), the term cartogram didn't appear until 1860, when it was used to describe small-scale maps of statistical data.

The word cartogram acquired the connotation of a purposeful non-conventional map projection in the 1960s after Waldo Tobler applied the mathematics of equal-area map projection to maps on which the size of areal units represented a transforming variable like population or wealth (Snyder, 1993).

Traditional thematic maps using a topographic map base often give a false impression of the spatial patterns of human geography, because many area units with large territories often have small populations, whereas relatively small units may have large populations. Not surprisingly, in some extreme cases, most of the patterns simply aren't visible.

One of the remedies to fix this is to insert an inset map to enlarge smaller units with large populations, which often leads to more design headaches about the layout, scale and interpretation. With more computer algorithms being implemented, cartograms now can be made with greater ease within and without a GIS (for more information, see *www.ncgia.ucsb.edu/projects/Cartogram_Central*).

Effective Spatial Structures

One of the defining characteristics of cartograms is their shocking unfamiliarity to map readers. Because a cartogram often uses a variable other than area as its base, it can reveal some spatial structures more effectively than traditional choropleth maps based on a topographic base.

I should also point out that this obvious advantage of cartograms also turns out to be a major impediment for their use, especially in the case of continuous cartograms, where the shape and orientation of units often are distorted.

Most people are excessively entrenched into the traditional topographic map base, but we shouldn't underestimate the inertia of map readers in adjusting to new views of their world. The encouraging news is that ongoing research has shown that most people actually can interpret cartograms in meaningful ways after getting basic explanations on how a cartogram is made.

In principle, the idea of a cartogram is similar to the creation of a homunculus: a representation of the human body in which each part is scaled in proportion to the size of the brain region devoted to it. To most people, a homunculus gives a rather distorted view of the human body, but from a neurological science perspective, it presents a more realistic and accurate picture of who we really are. A cartogram is essentially the geographical equivalent of a homunculus.

Cartograms of Politics and Economics

If you're still not convinced by the utility of cartograms, I'd like to call attention to two illustrative examples. One is the cartograms about the results of the 2004 U.S. presidential election obtained with the Gastner and Newman algorithm at state and county levels (*www.personal.umich.edu/~mejn/election*). It doesn't take a particle physicist to realize how close the election was on the cartogram, whereas the choropleth maps (used by most news media reporting the election) gave readers the false impression that George W. Bush had a landslide victory over John Kerry.

A second example is a cartogram that economist Jeffrey Sachs (2005) used in his recent article in *Scientific American*. Like in the previous case, Sachs' world-poverty cartogram presented readers with a vivid illustration of the enlarging gap between "Haves" and "Have-nots" at a global level. The work of a world-class economist represented in a cartogram allows us to conclude that something needs to be done about the dire situation of the extremely poor.

Cartograms as Necessity

In his presidential address to the Royal Geographical Society in 1962, Sir Dudley Stamp remarked that "the fundamental tool for geographical analysis is undoubtedly the map or, perhaps more correctly, the cartogram." Echoing Stamp's remarks 30 years later, Danny Dorling (1994) further argued that "cartograms should not be seen as just another option in a cartographic toolbox, but as a fundamental necessity in the just mapping of spatial social structures."

Indeed, colleagues in other disciplines, such as physicist Michael Newman or economist Jeffrey Sachs, have recognized such necessity as demonstrated by their brilliant work in using cartograms. Isn't it time for GIS users to take cartograms more seriously?

(This column originally appeared in the December 2005 issue of GeoWorld.*)*

The Yora of the Amazon and the Power of a Map

By Nigel Waters

Sitting on my bookshelf is a piece of history: the 1972 copy of the *Cambridge University Expeditions Journal*. In 1971-1972, I was the secretary to the Cambridge University Explorers and Travellers Club. The society's mission was to provide a weekly series of talks given by famous explorers as well as support a dozen or more student-led expeditions around the world.

On returning to Cambridge in the fall, each expedition submitted a report that was subsequently published in that year's journal (and now on the Web at *www.srcf.ucam.org/cuex/past.htm*). I was reminded of my "earlier life" when reading Alun Anderson's interview with Conrad Feather that appeared recently in *New Scientist* (see "People of the Forest," Vol. 182, No. 2449, pages 46-49).

A Feather in Peru

Feather was the first student at St. Andrews University in Scotland to win that university's Prize for the Environment, an annual award to individuals who develop innovative solutions to environmental problems by combining good science with economic and political realism.

Feather won the prize for his work with the Yora in southeastern Peru. Feather and four undergraduates from Cambridge began working with the Yora in April 2000, and their expedition was supported by the Cambridge University Expeditions Society.

Although the *New Scientist* article refers to this indigenous group as the Nahua, Feather prefers the less pejorative term Yora, because Nahua means "outsiders" or "foreigners," a common designation for marginalized groups ever since the Anglo-Saxons first referred to the Welsh that way in the sixth century.

No Summer Vacation

In April 2000, Feather put together his research team and headed out on a reconnaissance trip to the land of the Yora, whose home lies on the western borders of Manu National Park in southeastern Peru (*whc.unesco.org/sites/402.htm*).

Until the 1980s, the Yora had resisted all contact with civilization, but then they had to fight against the loggers who wanted to exploit their lands. With contact and hostilities came epidemics of disease, and the Yora died by the hundreds. Only 250 survive today, and they have all retreated to the village of Serjali.

Feather's reconnaissance trip was a success, and he and his interpreter, Frederico, and the rest of his research team were invited back. The Yora realized that they had a friend from the outside world that would use his knowledge to fight for their rights against the encroaching loggers.

Feather's group informed the Yora that their lands were protected as a reserve. The Yora were told of their rights, and all their meetings were recorded, in Spanish, in a "book of acts."

Feather left to finish his degree in Cambridge, but the loggers returned in force the following summer. The Yora left their land and headed for the nearest friendly town and won the support of the local authorities.

The "book of acts" was their most potent weapon in attempting to establish their rights to the land and the illegality of the loggers' acts. Feather returned and spent much of the next three years learning Spanish and Nahua as well as working for the rights of the Yora.

The Power of a Map

The Yora want legal title to the land they inhabit, which, at more than 2,000 square kilometers, is 10 times the largest land claim ever awarded to an indigenous group in Peru. How could they win the land? They could make a map of it.

For the Yora, every place had a name and a meaning. Feather and company explained that the government wouldn't understand their names. Also, they would have to use Global Positioning System (GPS) devices to give every place a number, and these numbers then could be used to draw a map.

The Yora were shown how to use GPS devices. According to Feather, "they got it straight away." They spent the next year building the cultural map and storing it in the computer. The lands where they hunted and fished and buried their dead were all duly marked. This is their land—their home—and the map showed how every part was special to them.

The Yora want this large area because it was theirs before, and they live a subsistence way of life that requires large tracts of land to support small populations. The Yora argue that they, and they alone, can act as effective guardians against the loggers.

Web Site Work

To promote their work, Feather and others founded the Peru-based non-government organization, Shinai Serjali, which has its own Web site at *www.serjali.org*. The organization's mission statement is provided on the site: "To support isolated indigenous peoples who live in the remote rainforests of Southeast Peru. We are helping them defend their rights, land and resources in order to provide a secure and healthy future for their children and help protect biodiversity."

Feather wants to use some of his prize money to donate a laptop with solar panels, so the Yora can continue to add information to the map. It's the "old idea" that "if you give someone a fish, they will eat for a day, but if you teach them to fish, they will eat forever." Now the Yora can use GPS devices, digital cameras and e-mail from the nearest town. Soon they should be able to protect their special places themselves.

That's Feather's goal. He wants to become redundant. After the Yora have the tools to map their environment and protect their interests, he will move on. Although I get the impression that he will come back to visit.

Serjali Thesis

Feather's undergraduate thesis is an interesting document, demonstrating an almost unbearable tension between his postmodern, undergraduate upbringing and the uses he found for his "kit bag" of high-tech tools. Even more of an intellectual struggle is his wish to intervene in the Yora's fight. He believes that the Yora have a right to their own destiny.

On reading all the material relating to Feather's work, I felt a relentless curiosity as to whether he actually made any use of GIS. He had GPS units and computers, and he was producing maps, but was he doing any GIS?

I e-mailed him, and he was kind enough to reply. He said they had made some use of ESRI's ArcView, but a donated copy of Avenza's Map Publisher was more useful for producing the maps on which the land claims were based, because it gave greater map design flexibility. He also noted that the Garmin 76 GPS unit was a particular favorite with the Yora, because it floated nicely when dropped in the rivers.

The Debate

The Yora's fate includes some controversy. The loggers want to exploit the land for its resources, but the Yora themselves also want to log, and they want to trade for modern goods that would make their lives easier. They would rather hunt with a gun than a bow.

Anthropologists are part of the debate, arguing that there may be no real pristine communities that haven't evolved through at least some interactions with the outside world. According to Thomas Headland, the 3,500-member Kayapo nation of Brazil is no longer interested in acting as conservationists of their land (see "Revisionism in Ecological Anthropology," *Current Anthropology*, Vol. 38, No. 4, pages 605-609). Conservationists also have argued against allowing indigenous people to exploit their own resources.

My recent GIS work with my graduate students has involved mapping traditional knowledge for the Tulita community, an indigenous people located not in the steamy Amazon, but in Canada's frozen north between Inuvik and Yellowknife. This information will be placed in a GIS and used to protect important places from exploitation.

But what if the Tulita themselves want to exploit their lands for economic purposes? In my view, that's their right, but it's also my goal to work with them and help them sustain their environment.

(This column originally appeared in the December 2002 issue of GeoWorld.*)*

What Can GIScientists Learn from Marshall Islanders?

By Daniel Sui

During the recent holidays, many of my friends shared their excitement about using the latest online version of Google Earth for planning travel activities. This is further evidence that the number of technological innovations for spatial data handling at the personal level during the last two decades has been tremendous.

As Matt Ball, *GeoWorld* editor, perceptively observed (see "Google Verifies that GIS Is Media," December 2005, page 6), the development of Google Earth has validated the idea of GIS as media, a notion that Mike Goodchild and I speculated on about five years ago.

Retrieving Stick Charts

Marshall McLuhan argued in *Laws of the Media* that any innovations in the dominant mode of communication media would simultaneously enhance, obsolesce, retrieve and reverse some aspects of social practices. I won't discuss all of the implications of McLuhan's laws for GIS, but this column describes some of my recent thoughts on the "retrieval" aspects of media transformation, which argues that certain social or cultural practices pushed aside long ago can eventually be retrieved in light of new technological innovations. In the context of GIS, retrieval means a renewed interest in earlier ways of handling geospatial information.

Among the many ways of dealing with geospatial information throughout history, I have been most impressed by the stick charts that Marshall Islanders used for navigation. And I believe we still have a lot to learn from Marshall Islanders and their unique ways to handle spatial information.

The Marshall Islands consist of two parallel island chains, Ratak (eastern) and Ralik (western). The low elevation of the islands, set at right angles to the ocean swells created by dominant trade winds, and the long distances between atolls make them particularly difficult to glimpse from the sea. In the Marshall Islands, orientation and navigation are based on wave patterns. With the help of stick charts, trained islanders skillfully read the waves and watch for certain swells that show the direction of land (see Plate 5).

Stick charts are unique products of the local environment. They're often constructed by palm ribs or pandanus wood bound by coconut sennit with cowrie shells added to represent the islands. There are three basic types of stick charts in the Marshall Islands: mattang, meddo and rebbilib.

The mattang, also known as wabebe, are used only for instruction to show the novice fundamentals about navigation. Seamlessly woven into the stick charts is the idea of wave piloting. Straight sticks are used as a framework, and curved sticks indicate ocean swells curving from contact with islands. To teach the recognition of wave patterns at sea, young apprentices undergo practical training, and all navigators have stick charts on which the orientation and sequence of the currents and wave patterns are marked.

The second form of stick charts, meddo (medo), is an extension of the mattang. They show swell patterns in relation to a number of islands or a group of atolls. In this respect, the chart is an extension of the islands' swell patterns. The function of the meddo is to show the position of the islands relative to observable swell phenomena. In general, real distances and directions aren't the primary concern.

The final type of stick chart is the rebbilib (rebillit), which cover the whole (or a large part) of the archipelago, and they're concerned more with islands than with patterns of swells. A rebbilib normally covers an entire chain of islands or atolls with a clear illustration of the spatial configurations of the islands represented.

A rebbilib is closest to what we consider a map. Some cartographic historians have argued that Western mapping techniques heavily influenced rebbilibs. Among the three types of stick charts, only the mattang is considered to originate from the Marshall Islands.

Examples of each type of stick map can be found online at *www.janesoceania.com/micronesian_stick_chart*. If rebbilib is an analog GIS, meddo may be roughly compared to the components of a GIS database, and mattang is the approximate equivalent to metadata.

Sticks and GIScience

Although they may appear simple, crude and idiosyncratic compared to current technologies, stick charts are quite sophisticated. Marcia Ascher (2002), for example, evaluated the mathematical accomplishments demonstrated by stick charts. I also believe that stick charts can teach us useful lessons about many fundamental issues in GIScience.

First, stick charts provide an interesting and alternative way to represent space. Instead of the Cartesian coordinate system, in which location and attribute information are separately encoded through an object or field view, stick charts represent multiple locational and attribute information holistically. As features of the land, sea and winds are placed in a planar representation using simple devices, a relational view of space is embedded in stick charts.

Second, stick charts represent a relational view of space as well as a flexible model about how the world works. Unlike a typical map or GIS, stick charts don't depict islands in fixed geographic relationships. The relative position of the islands in regard to prevailing winds and ocean currents/swells is emphasized, which often is ignored on conventional charts.

Third, stick charts have high-level, built-in flexibility. Although stick charts, like paper maps, play a dual role of storage and display medium, the shells on a stick chart can be adjusted to represent any islands. The sticks also can be re-orientated at multiple angles to meet special needs or circumstances. Stick charts introduced the concept of interoperability.

In addition, the application of stick charts is a communal, participatory process. To make stick charts useful, users were trained to understand diverse natural processes, such as weather conditions, wave movements in the ocean, etc.

The use of rojen Koklol (navigation formulas) and alinlokonwa (sailing songs) also were stressed in the training process. The incorporation of time during a trip, for example, is based on the intervals of sailing songs. It's amazing that much of the training was done through elaborate oral traditions—there was no written language among Polynesian, Micronesian or Melanesian peoples.

All of the information on stick charts is memorized, and the charts themselves aren't actually taken on journeys, but used as memory or learning aids to rehearse wave patterns to be encountered. This is significant, because the islanders recognized the limitations of the stick charts, so they focused their attention on the ends, while treating the devices as mere means to those ends.

David Turnbull (2000) argued that the common element in all knowledge systems (including western scientific knowledge) is their "localness" and contingent nature. Thus, each system has utilities as well as limitations. The great divide between scientific and indigenous knowledge turns out to be dangerously misguided, according to recent research.

Humanistic GIScience

I commend recent efforts by researchers to link GIS with indigenous knowledge for theoretical and application purposes. I consider these efforts to be evidence of retrieval, according to McLuhan's law of media. This is a wide-open area for further research under the general umbrella of humanistic GIScience.

More than 30 years ago, Fritjof Capra revealed the parallels between modern physics and eastern mysticism in *The Tao of Physics*. Capra demonstrated that what's often perceived as crude, primitive and "mystical" often shares astonishing similarities with "cutting-edge" scientific discoveries.

The recent efforts by the scientific community in general, and GIScientists in particular, to link indigenous knowledge with recent scientific and technological innovations demonstrate that there's a lot for us to learn from people in other places and times.

In the case of geospatial information, we're still struggling with what Marshall Islanders achieved using simple sticks and cowrie shells, although we're using powerful computers. Furthermore, the stick charts used by Marshall Islanders may encapsulate a robust alternative conceptualization of the world that's needed for the "next generation" of geospatial technologies.

To learn how other people have perceived and represented their worlds in an efficient way may be one of the most important lessons that geography can teach us. As psychologist Richard Nisbett (2003) argued, the most interesting geography is actually the geography of thought.

(This column originally appeared in the March 2006 issue of GeoWorld.*)*

Section 3. References

Alexander, S., Callaghan, C., Paquet, P. Waters, N. 1996. "A GIS Predictive Model for Wolf Habitat Use in Peter Lougheed Provincial Park, Alberta." *GIS '96 Conference Proceedings*, GeoTec Media, Boulder, Colo.

Ascher, M. 2002. *Mathematics Elsewhere: An Exploration of Ideas Across Cultures*. Princeton University Press, Princeton, N.J.

Batty, M., Slack, A. 2002. "Virtuality and Cities: Definitions, Geographies, Designs," *Virtual Reality in Geography*, Fisher, P., Unwin, D. (eds.), Taylor & Francis, New York.

Cheng, Y., Hodgson, M.E. 1996. "Rectification and Registration of Digital Imagery by Using Control Points and Control Lines," *Integrated Computing Applications Section*, Computational Physics and Engineering Division, Oak Ridge National Laboratory.

Clarke, K.C. 2001. *Getting Started with Geographic Information Systems* (Third Edition). Prentice Hall, Upper Saddle River, N.J.

Dobson, J.E. 1992. "Spatial Logic in Paleogeography and the Explanation of Continental Drift," *Annals of the Association of American Geographers*, Vol. 82, No. 2, pp. 187-206.

Dobson, J.E. 1996. "A Paleogeographic Link Between Australia and Eastern North America: A New England Connection?" *Journal of Biogeography*, Vol. 23, No.3, pp. 609-617.

Dobson, J.E. 1998. "The Iodine Factor in Health and Evolution," Geographical Review, Vol. 88, No. 1, pp. 1-28.

Dorling, D. 1994. "Cartograms for Visualizing Human Geography," In Hearnshaw, H.M., Unwin, D.J. (eds.), *Visualization in Geographical Information Systems*, pp. 85-102. John Wiley & Sons, New York.

Evink, G.L., Garrett, P., Ziegler, D. (eds.). 1999. *Proceedings of the Third International Conference on Wildlife Ecology and Transportation*. FL-ER-73-99. Florida Department of Transportation, Tallahassee, Fla.

Gastner, M.T., Newman, M.E.J. 2004. "Diffusion-based Method for Producing Density-equalizing Maps." *Proceedings of the National Academy of Sciences*, Vol. 101, pp. 7499-7504.

Goodchild, M.F., Steyaert, L.T. 1996. *GIS and Environmental Modeling: Progress and Research Issues*. GeoTec Media, Boulder, Colo.

Goodchild, M.F., Parks, B.O., Steyaert, L.T. 1993. *Environmental Modeling with GIS*. Oxford University Press, New York.

Herrero, S. 1985. *Bear Attacks: Their Cases and Avoidance*. Hurtig Publishers Ltd., Toronto.

Krakauer, J. 1995. "Rocky Times for Banff." *National Geographic*, Vol. 188, No. 1, pp. 46-69.

Nisbett, R.E. 2003. *The Geography of Thought: How Asians and Westerners Think Differently...and Why*. Free Press, New York.

Pole, G. 1998. *Healy Park*. Mountain Vision, Field, British Columbia, Canada.

Raisz, E. 1938. *General Cartography*. McGraw-Hill, New York.

Robinson, A.H. 1982. *Early Thematic Mapping in the History of Cartography*. University of Chicago Press, Chicago.

Romm, J. 1994. "A New Forerunner for Continental Drift." *Nature*, Vol. 367, pp. 407-408.

Sachs, J. 2005. "Can Extreme Poverty Be Eliminated?" *Scientific American*, Vol. 293, No. 3, pp. 56-65.

Salem, L., Testard, F., Salem, C. 1992. *The Most Beautiful Mathematical Formulas*. John Wiley, New York.

Simon, H. 1991. *Models of My Life*. Basic Books, New York.

Snyder, J. P. 1987. *Map Projections—A Working Manual*. U.S. Geological Survey Professional Paper 1395. U.S. Government Printing Office, Washington, D.C.

Snyder, J.P. 1993. *Flattening the Earth: Two Thousand Years of Map Projections*. Chicago, The University of Chicago Press.

Turnbull, D. 2000. *Tricksters and Cartographers: Comparative Studies in the Sociology of Scientific and Indigenous Knowledge*. Taylor & Francis, Oxford, England.

Wegener, A. 1912. "Die Entstehung der Kontinente." *Geologische Rundschau*, Vol. 3, No. 4, pp. 276-292.

Wright, J.K. 1938. "Problems in Population Mapping," *Notes on Statistical Mapping, with Special Reference to the Mapping of Population Phenomenon*, pp. 1-18. American Geographical Society and Population Association of America, New York.

Section 4.
GIS and Hot-Button Issues

Introduction
By Jeremy Crampton

The columns in this section share a common theme: technology has social consequences. Whether it be the conflict between environmentalists or Dobson's column written in the immediate aftermath of the 9/11 attacks, the columns share a perspective that what's important about GIS is not the technology itself (something they spend little time on). Rather, the columns focus on how GIS is applied, used, and the effects of that deployment. No man is an island, and neither is technology—it is intricately embedded in society, and can also profoundly affect it.

Several of the columns speak to each other across the years. When I quoted Neil Smith's famous aphorism that the 1992 Gulf War "was the first full-scale GIS war," I intended to issue a warning to what I saw was the too complacent endorsement of militaristic uses of GIS. Compare then Waters' column from December 1992 a dozen years earlier who used the same quote. For Waters, while these militaristic uses of GIS exist, he also points to more positive peacetime uses, many of which he sees as deriving from Pentagon funding or affiliated organizations such as the RAND Corporation.

Many of Dobson's columns address related issues of how GIS can be turned to more positive uses. In one of the columns reproduced here, Dobson recounts how he visited Ground Zero in New York City, shortly after 9/11, when an Airbus A300 crashed on Long Island near his old home. He observed the GIS team swing into action, very quickly producing maps and analysis. Dobson gained a very personal perspective that day on the work of these volunteers.

Dobson also details the work of the Landscan project, which he used to head. This project seeks a more accurate and realistic assessment of where people are not just at night but during the day (the U.S. Census had 55 people in the block around the World Trade Center, an obviously unrealistic number). This project actually points to one of the deepest problems facing social science today: the opposition between continuous and discontinuous information. While the world is lived "continuously" as it were (eg., differences between humans vary continuously across space), we often use methods that are discontinuous or discrete (eg., the choropleth map). Landscan aims to resolve this difficulty by resorting to an old but rarely used map, the dasymetric. It would be nice to see this map form more directly implemented in GIS.

On a final note, I wonder if, taken as a whole, we are not a bit too boosterish about

GIS. There's no doubt that GIS is useful in many circumstances, and that message does need to get out to an often geographically uninformed world. But I think there's also something to be said for a more critical attitude, one that constantly and productively questions received wisdom. I would point to the students, who made a number of positive suggestions for more deeply understanding the links between society and GIS (Crampton, "GIS Needs to be Part of the Solution"). In this way perhaps we can transform geography from "an anemic discipline steeped in trivia" as Dobson memorably puts it, to something more politically relevant.

GIS Helps Estimate Populations at Risk

By Jerry Dobson

A quiet revolution is taking place in population geography. GIS is helping build new databases, and they, in turn, are affecting all sorts of geographic applications. They're especially important in estimating populations at risk in natural disasters, technological accidents, terrorist attacks and regional conflicts. Advances have come at two levels: one offering global coverage and the other tackling the world one region at a time.

Global Population Geography

I've written before about the LandScan Global Population Database produced by Oak Ridge National Laboratory (*www.ornl.gov/sci/landscan/*). It's at 30-arc-second (less than 1-square-kilometer) resolution, updated annually and available to all comers for free.

But even finer resolutions are needed for many types of disasters, especially those that affect urban areas. Daytime and nighttime population estimates at 3-arc-second resolution (90 meters by 90 meters at the equator) are possible for the United States due to the nation's established collection of high-resolution spatial databases, but what about the rest of the world?

Global threats demand detailed population figures worldwide, but estimates for buildings or city blocks and daytime vs. nighttime estimates are out of the question for all but a few advanced nations. Faced with such overwhelming need and the impossibility of producing a suitable global database, I've worked for several years now to develop a technique that can be employed in real time after the location and extent of any actual or potential disaster is known.

The problem has been decomposed into two parts: 1) estimating how many people are typically in each type of building by day or night, and 2) measuring how much floor space is available for occupation. The second component—measuring floor space through field observation, map analysis or image analysis—is labor intensive and prohibitive in its worldwide application. It can be done for specific locations where the actual or potential disaster is known and the need is clear and compelling. The first part—estimating population density for various building types—must be done with forethought for large world regions.

Regional Approach

This updated population-estimation method is based on occupancy tables for various (approximately 39) building types, outdoor spaces and transportation lines. The database is continuously improved to develop a standard notion of how many people each building, vehicle or open space will hold, estimate fluctuations for night vs. day and provide standard estimates of people per area. The result is, in fact, a table of coefficients (density expressed as people per 10 square miles) for night and day that can be multiplied times the area after the area of a given building is known.

Standing on Shoulders

Both LandScan and the updated approach are prime examples of how valuable and enduring the legacies of geographic research can be, or, conversely, how deep current information technologies reach into geography's collective "kit bag." The global LandScan database, for example, heavily relies on dasymetric interpolation invented by John K. Wright of the American Geographical Society more than 65 years ago.

The regional approach is based on a revival and enhancement of methods used by settlement geographers principally from about 1920 to 1970. During that era, diverse geographical studies were made of building types, forms and styles. For example, Albert Demangeon (1872-1940) examined vernacular dwelling designs, sometimes including detailed floor plans (Demangeon 1930, 1937), and Amos Rapoport continued to conduct studies of house form and culture as late as 1969 (Rapoport 1969). Fred Kniffen (1901-1993) focused on characteristic folk styles of housing, barns, fences and outbuildings (Walker and Richardson, 1994). Kirk Stone (1914-1997) focused on the spatial organization of settlements (Dobson et al, 1999).

Stone actually tried to map the boundaries of continuous settlement for the whole Earth. Near the end of his life, when asked what was his greatest contribution, he claimed it was the continuous vs. discontinuous settlement concept. He lamented, "It seems not to have been very useful," then added prophetically, "but it might be in time."

Stone would be pleased to learn how important such measures have become in current efforts to counter global threats posed by nuclear, biological and chemical agents. Atmospheric transport models for predicting contaminant plumes of anthrax, for example, have improved remarkably during the last few decades, especially in terms of spatial resolution, and there's a pressing need for more precise population geography to identify people at risk. Only now are population and settlement distributions, including the fringes of settlement, being mapped at spatial resolutions suitable for intersecting with contaminant plumes.

(This column originally appeared in the November 2000 issue of GeoWorld.*)*

The Barrage of Baghdad Changes Everything

By Jerry Dobson

Sit down, my friend. It's time for some straight talk about geography and GIS. We geographers may chuckle politely when you tease us about knowing our "states and capitals," but we bombed Baghdad while the lights stayed on, and we can track you down like a dog. Now listen up, and nobody will get hurt.

That's what I'm tempted to say as I watch the news these days. GIS just changed warfare as profoundly as dynamite did in 1867 or aircraft in World War I or the atomic bomb in World War II. Meanwhile, back home, GIS is affecting just about everything that involves location, movement or flow, and GIS-based human-tracking devices threaten to redefine social relationships. Yet public ignorance of geography and GIS is so great that practically no one knows.

Surreal Silence

During the Iraq War, it was surreal to hear officials and reporters daily spout new acronyms from the battlefield (e.g., RPG, BDA, AAA, EPW) without once saying "GIS." They marveled at precision bombing, never mentioning that GIS is the foundation of it all. The closest they came was "GPS" (Global Positioning System), which demonstrates how little they understand the limitations of that particular technology.

Those in the geographical sciences didn't really bomb Baghdad, of course, but for thousands of years geographers, cartographers, geodesists, surveyors and, more recently, remote-sensing specialists advanced the science that made precision bombing possible. While society snoozed in blissful ignorance, we changed its fabric as fundamentally as chemists did in 1867 or physicists in 1945.

The difference this time is that no one knows what's behind it, and that's scary. Imagine a future in which GIS continues to progress at its rapid pace, while society stays ignorant of geography and GIS.

Worst of all, people won't know how to protect themselves from us. Why would they need to protect themselves from us? Well, it won't be you and me as individuals that they have to fear. It will be a powerful military/industrial/geographical complex that will go wherever the money leads. Even the "good guys" who pioneered commercial GIS will be swept aside in a flood of new technology, marketing and commercial opportunities. To turn this juggernaut around, investments in geographic education, training and safeguards must be comparable to those accorded to nuclear physics after World War II.

Geographic Literacy

This call may sound too bold, but we've been much too timid in the past. We've politely admonished society to pay attention to geography. At most, such efforts convinced some states to require geography courses in high-school curricula, and the people who made that happen are to be commended. To the public at large, however, we have presented an abysmal image of an anemic discipline steeped in trivia.

Every so often, a new survey tests the "geographic literacy" of teenagers, followed inevitably by lamentations that kids can't find [pick a] country or [pick a] sea on a world map. I'm as disappointed as anyone else with that type of "geographic illiteracy," but I cringe at the image of our discipline presented in those headlines.

What's disturbing to me and anyone else who understands geography is that the people who write those surveys don't seem to recognize geography as a science. Granted, it's difficult to compose a questionnaire that tests anything other than place-name geography, but surely someone could do better than all we've seen so far.

Why not devise a GIS-like exam that requires test takers to exercise analytical skills? The results, good or bad, would be a more realistic gauge of geographic knowledge than simply asking where places are. And the test itself would send a respectable message to the public about what geography entails.

No More "Business as Usual"

Public awareness of geography is growing—slowly, steadily, due mainly to the diffusion of GIS throughout society rather than our paltry public-relations efforts, but the barrage of Baghdad says it's time to kick it up a notch. Society can't afford to continue "business as usual" with geography and GIS anymore than it could do so with nuclear physics and engineering after Hiroshima.

Precision-guided weapons—G-bombs (G for geographic) might be a better name—change everything, just as the A-bomb did. And the same technology that guides a bomb can guide a child, wife, employee or slave. Safeguards are essential.

Warfare always shakes the very foundations of society. [Update: Already, this war's remarkable technology is eroding privacy and controlling people. More that 50,000 U. S. workers are "geofenced" by GPS tracking devices as part of a vast social experiment that John Pickles calls "the surveillant society," Harlan Onsrud calls, "the tragedy of the information commons," Dan Sui calls "your stolen geography," and I call "geoslavery."] The whirlwind of progress in both directions, warfare and geoslavery cries out for public debate, but there has been none. Why? Because society doesn't understand geography, the intellectual home of GIS.

So listen up, my friend. Get serious about geography. Pay attention to GIS. See what we've done to warfare, politics and international relations. See how we affect your life each and every day. Promote the good. Oppose the bad. Insist on public debate. Act soon, my friend, or we'll all pay the price.

(This column originally appeared in the July 2003 issue of GeoWorld.*)*

Rethinking GIS and (Homeland) Security

By Jeremy Crampton

"The war against Iraq in 1990-91 was the first full-scale GIS war."
—Neil Smith, 1992

Smith's now (in)famous statement was the opening line of an article examining the links between GIS and war—specifically in the first Gulf War. Smith was troubled that GIS—along with many modern weapon systems—enables us to forget the actual effects of war by substituting a slick, scientific virtual reality. By using "geosmart" bombs, the Global Positioning System and 3-D representations on screens well away from the "sharp end of things," it becomes easier to wage war.

If Smith is right about GIS making war more "doable," then today we're living through a second GIS war. But it's occurring in a way that Smith didn't predict: through "security."

GIS and Security

"Security," or, more precisely, homeland security, now is American law and official policy. These developments represent important shifts in American policy and mark a new political era that abandons the Cold War's policy of containment and deterrence for one of threat assessment and unilateral action.

The great Prussian military strategist Carl von Clausewitz observed that "war is the continuation of politics by other means." In other words, if you engage in war, you should do it to achieve your political goals (and not just for self-defense, ambition or because it's a "just" war). Today, von Clausewitz would have a new slogan: "security is the continuation of politics by other means."

As with the first Gulf War, the war on terrorism and the methods of security designed to prosecute that war rely heavily on GIS and cartographic technologies. The relevance of mapping and GIS was demonstrated soon after the terrorist attacks on Sept. 11, 2001, when the New York Times and NBC published several powerful light detection and ranging (LIDAR) images by The Center for the Analysis and Research of Spatial Information at Hunter College. The LIDAR images showed a "before" and "after" view of the World Trade Centers in 3-D. In that first week, LIDAR images were one of the few ways to penetrate the smoke and dust.

In the following months, ESRI offered a series of seminars around the country on how GIS could assist in emergency prevention and response, published white papers, produced a CD-ROM on security, and established a Web site for GIS and security. The Association of American Geographers helped publish *The Geographical Dimensions of Terrorism* (Cutter et al., 2003) that sees "geospatial data and technologies infrastructure research" as a top priority. Such recognition of GIS is based on the Department of Homeland Security's Information Analysis and Infrastructure Protection Directorate's budget request for "development and maintenance of a complete and accurate mapping of the Nation's critical infrastructure and key assets."

Second Thoughts on Security?

These developments (and many others) in the name of security are neither surprising nor totally desirable. They come with a downside, as Mark Monmonier's (2002) eye-opening book, *Spying with Maps—Surveillance Technologies and the Future of Privacy*, clearly demonstrates. Monmonier presents a startling array of technologies that compromise our ability to be private in our own locations. Our movements through space now can be tracked completely and comprehensively without our knowledge or permission. Much of this is possible through the very "geospatial data technologies" lauded in *The Geographical Dimensions of Terrorism*.

These developments are taking place without any debate or even recognition among GIS users or the public about this political use of GIS and security. What can we (GIS users and the public) do? The following are some initial suggestions:

1. We need to be less complacent about becoming a research arm of the security industry.

2. We need to recognize that GIS technology isn't an application-independent tool and that it can be politicized.

3. We need to challenge the aura of paranoia and threat that some policy makers and geographers instill.

4. Finally, we should develop GIS and mapping capabilities that are richly meaningful to our human experience rather than ones that promote racial, class or social divisiveness.

(This column originally appeared in the April 2004 issue of GeoWorld.*)*

GIS Supports the War on Terrorism in Many Ways

By Jerry Dobson

Since the terrible events of Sept. 11, 2001, I've been able to watch how GIS has been used to help fight the war on terrorism. The following are three examples—one as a firsthand observer, one at a distance and one as a contributor.

American Flight 587

On Nov. 12, 2001, I visited New York City's Emergency Operations Center on Pier 92. I came to see the GIS facility that was set up so urgently in the aftermath of Sept. 11. At about 9:25 a.m., I noticed a subtle flurry of activity in the vast open hall—nothing dramatic, just a few policemen walking briskly toward the Command Center. Then, someone rushed over to announce that a plane had crashed in the Rockaways. Eight GIS team members were on duty at the time, and everybody kicked into action. Within 30 minutes, the first map was available to emergency responders. It contained high-resolution images of the neighborhood with every street named and every house clearly identifiable.

That's a remarkable accomplishment—especially impressive to me, having seen firsthand how much took place in that brief time. Initially, no one knew what size plane was involved or precisely where it had crashed. The first report said the plane was in the water, as indeed the tail section was. No one knew, of course, whether the crash was an accident or another terrorist attack, but it surely sounded like what had happened one month earlier. Prudently, the city closed all bridges and tunnels for most of the day.

Minutes later, someone from the Command Center announced a specific street name (Beach 129th St.) and then the intersecting street (Newport Ave.). That turned out to be a couple of blocks off, but the damage zone and map were large enough that it didn't matter. About 20 minutes after the first alarm, I called my wife, Gwen, back at the hotel. While we spoke, we heard the announcement first that it was a large plane, then that it was a Boeing 737 (actually an Airbus A300). Through it all, police and fire responders were insistent on getting good maps, a lesson brought home to many of them during their rescue and recovery efforts at the World Trade Center.

Just how far had the GIS facility come since Sept. 11? The original GIS was housed in Building 7 World Trade Center and perished in the attack. Before that day, New York City had NYCMap (pronounced "nice map"), one of the most exquisite municipal geographical databases in the world, amounting to a fine-resolution, 3-D model of the city. Suddenly the city's system was gone, but, fortunately, one copy remained with the geographers who developed it at Hunter College, City University of New York. Sean Ahearn literally unplugged his own computers, loaded them in a van, and drove to the Command Center. A call went out through GISMO, the metropolitan area GIS users group. About 50 GIS analysts responded, and they worked around the clock to reestablish the system. Operations were shifting to a permanent location elsewhere when I came to visit a month later, but 37 maps of Ground Zero still adorned the walls as testament to their accomplishment. Policemen and firemen had learned to call those maps by number. The crash of American flight 587 proved conclusively that the temporary system and its permanent counterpart were ready for whatever might next befall the "Big Apple."

Knowing that volunteers had been called before, I hung around for a while to see if I could be of service. Mustering the accumulated skill of 26 years of GIS experience, I deftly sidestepped the real performers and masterfully kept out of their way. They performed flawlessly, and it soon became clear that no outside help was needed. Later that evening, Gwen and I left for home on one of the first flights permitted out of JFK Airport.

The Anthrax Investigation

Months before the terrorist attack, I was invited to speak at the Centers for Disease Control and Prevention (CDC) in Atlanta. The talk, scheduled for Oct. 19, 2001, was to focus on high-resolution population data and potential applications in epidemiology. By that time, domestic incidents of bio terrorism had given new urgency to existing and proposed data. Ultimately, the scope expanded to a broad discussion of CDC's plans to improve its agency wide GIS capability to meet the current crisis. I returned for more discussion on Oct. 22. Days later, my host, Jerry Curtis, a GIS analyst at CDC, was assigned to investigate the Brentwood Postal Facility, central to the anthrax investigation in Washington, D.C. Throughout the investigation, we spoke or exchanged e-mail frequently concerning technical matters. Curtis faced the vexing administrative and technical problems inherent in any new GIS endeavor, but he had to resolve them in days rather than weeks or months.

At CDC, I displayed a 3-D model of Washington, D.C., based on digital orthophoto quads at one-meter resolution draped over terrain and buildings. A consortium of companies and agencies, coordinated by the National Capital Planning Commission, developed the model. I used it, along with an air dispersion model, to illustrate what might be possible with sufficient resources to model the spread of nuclear, biological and chemical agents, and identify populations at risk. But that's not what Curtis had in the field. When the chips were down, he and ESRI Inc.'s Jay Devasundaram had to scan hard-copy floor plans. Eventually, they produced a precise spatial database of the facility, indicating positive cases and samples, work locations and break areas. The final version, displayed during President George W. Bush's visit to CDC, still shows the handwritten notes that marred the best available copy. That, my friends, is the real world of crisis epidemiology.

The Afghan Refugee Crisis

From its beginning in 1997, the LandScan Global Population Database (Dobson et al 2000, Dobson et al 2003) was designed to predict populations at risk from any type of threat, especially bio-terrorism, and there seems to be no end to the potential applications. As tension mounted in Afghanistan, inquiries focused on how accurate the database might be for that country. Fortunately, we were able to speak with confidence because virtually all of the country had been covered by the best global databases available, and Chris Auricht of the United Nations' Food and Agriculture Organization had conducted extensive field testing of LandScan data there long before the crisis began. Later, I was gratified to learn that our data are being used to plan the seed supply for the spring planting in that troubled region.

Accurate estimates of nighttime and daytime populations are essential for emergency response and recovery efforts, and they also could play an essential role in epidemiology. Recently we proposed an exceptionally precise version of LandScan for the United States that would show daytime as well as nighttime populations at "city block" detail for the whole countryside, and we demonstrated it for a 29-county area in Texas and Louisiana (Dobson 2003; Dobson et al 2003). On Sept. 11, we surely wished we had chosen New York City, instead of Houston, as our study area.

To illustrate the need for such estimates, consider that the United States Census 2000 showed only 55 people in the large Census Block that contained the World Trade Center, while the number of people there on Sept. 11, 2001, was at least 50,000 and perhaps much greater. Official census data generally record residential (nighttime) populations, but daytime populations (workers, students, shoppers, etc.) are potential victims in terrorist attacks, natural disasters and technological accidents. In our prototype study, we estimated 160,000 additional people in downtown Houston on a typical workday.

Outside city centers, official Census Blocks are too large to provide adequate spatial detail. Along with Ed Bright, Phil Coleman, Budhu Bhaduri of Oak Ridge National Laboratory and colleagues here at the University of Kansas, I remain committed to building such a database for the United States.

[Update: I led the LandScan Global Population Project at Oak Ridge National Laboratory (ORNL) from its inception in 1997 and LandScan USA from its inception in 2000, until I moved to the University of Kansas in 2001. I no longer have any connection with either project. Annual updates of LandScan Global are available to everyone for free on the ORNL Web site (*www.ornl.gov/sci/landscan/index.html*), and no patent is in effect. Since my departure, LandScan USA has produced daytime and nighttime population distributions at 3 arc-second resolution for most of the United States, but LandScan USA has not been released by the Department of Homeland Security.]

(This column originally appeared in the January 2002 issue of GeoWorld.*)*

How Will GIS Help Fight the War on Terrorism?

By Jerry Dobson

What role will GIS play in the world's new war on terrorism? Its contributions will be vast, no doubt, and the GIS community can take tremendous pride in the efforts already under way.

One week after the scurrilous attack, NBC television reporters toured New York City's Emergency Operations Center. The highlight of the segment that aired on "Today" and "Nightly News" was a fresh, airborne light detection and ranging (LIDAR) image of "Ground Zero" compared to the NYCMap (pronounced "nice map"), a pre-existing 3-D geovisualization of the World Trade Center based on high-resolution digital orthophoto quads draped over building heights.

Beside the monitor, describing the changed urban landscape, sat Sean Ahearn, GIS and remote sensing specialist from Hunter College, City University of New York. The Hunter team, like so many first responders, worked tirelessly after the attack, and EarthData International flew LIDAR missions daily from Sept. 15 onward. Their contribution to the rescue and recovery effort was significant. Their work supported searchers in vital ways that made onsite personnel safer and more effective, and their contribution surely will grow as New York City rebuilds.

One of Our Own

The TV reports didn't identify Ahearn. I recognized his face and voice, but to most viewers, he was just another nameless hero among the thousands who materialized on Sept. 11, 2001, and the following days. Later, an article in The New York Times named Ahearn and described the NYCMap, the LIDAR images and the technologies behind them (Chang, 2001).

Ahearn is one of our own in the GIS community, and we should know him well. He's an associate professor in the Geography Department at Hunter as well as director of the Center for the Analysis and Research of Spatial Information (CARSI). He holds a Ph.D. in environmental monitoring from the University of Wisconsin—Madison, and his specialties are geographic information science, remote sensing, digital image processing, natural resources and habitat studies. He and Jeff Osleeb, associate director of CARSI, have been active in the University Consortium for Geographic Information Science, and Osleeb is a Councilor of the American

Geographical Society (AGS). In February 2001, when the Association of American Geographers met in New York, CARSI hosted an open house at the "Mapping New York" exhibit to highlight the NYCMap (Dunlap, 2001).

CARSI was established to promote interdisciplinary research and teaching in geographic information science with a focus on the urban environment of New York City and its surrounding municipalities. It's a focal point for the collection, archival and analysis of geographic information relevant to New York. All this was in place well before the bombing, and the investment paid off in spades when the need arose.

Be Ready

As the new global alliance begins its worldwide campaign against terrorism, will geographers be called on as in previous wars? During World War II, one-third of all academic geographers moved to Washington, D.C., to work in the Office of Strategic Services (OSS) and other federal agencies. During World War I, AGS ran the "Inquiry," a multidisciplinary information-gathering, scientific-integration and geographic-analysis venture that was commissioned personally by President Woodrow Wilson. Clearly, the Inquiry, led by Isaiah Bowman, foreshadowed Bill Donovan's OSS.

This time, the needs will surround us wherever we may be. Everyone, everywhere must be vigilant, but the war itself won't require large numbers of people if it's managed properly. Undoubtedly, GIS will be called on as never before, and the expertise of geographers, computer scientists and related disciplines will be essential. But the numbers moving to Washington or elsewhere recruited into the war effort won't be as great as in previous wars. That's because existing government institutions are far more capable than at the beginning of any previous conflict, special operations will be favored over territorial campaigns and the civilian targets threatened by terrorists are ubiquitous.

CARSI was ready to aid New York City, but no one knows which city, county, state or country will be next.

Global Threats to Local Places

Terrorists can attack anywhere, but some places are more likely targets. Can "GIScience" help predict where? Yes, but some threats are more predictable than others. We can't be lulled into focusing exclusively on the type of attack that occurred in New York City, Washington, D.C., and rural Pennsylvania. Hijacked aircraft pose a continuing threat, but so do conventional weapons. Airport security will mean absolutely nothing, for instance, to a lone terrorist with a surface-to-air missile. That's most likely to occur in definable zones around airports and particular sites within those zones that can best be found with GIS.

That's prevention, but attacks of one sort or another—guns, germs or bombs—inevitably will occur. When they do, GIS will be essential to support emergency management.

For many years, the greatest fear among counter-terrorism insiders has been weapons of mass destruction, including chemical, biological and nuclear threats. The U.S. military's Hazard Prediction and Assessment Capability was built to predict where such plumes will go, and the LandScan Global Population Project (*http://sedac.ciesin.columbia.edu/plue/gpw/landscan/*) that I led at Oak Ridge National Laboratory was designed primarily to identify populations at risk during such attacks. One great need will be to distinguish whether an epidemic is natural or human-induced, and the best hint will be its geographic pattern of spread. GIS will be essential to the investigation.

In every instance, retaliation will be required, and no one can predict its unintended consequences. Collateral casualties are a continual concern, and GIS has been instrumental in producing the global population database that will be needed to minimize casualties—it's impossible to avoid hurting people if you don't know where they are. LandScan2000 is the most precise estimate yet as to where Earth's 6 billion people are.

But people don't stay put when the fighting starts. The first actual effect of the retaliation for the recent attacks isn't battlefield casualties, but rather a refugee crisis of tragic proportions. A million or more innocent Afghans are on the move, and they will face hunger, thirst, disease and hostility even without any military strikes. Here again, as demonstrated in Kosovo, LandScan2000 can serve as a baseline for monitoring refugee movements, and GIS can serve as a spatial accounting tool for reconciling diverse observations.

Ultimately, an even greater fear is that retaliation will spark regional conflicts or widespread war. If that happens, all bets are off, and the GIS community will be mobilized as never before. Indeed, if the alliance really needs large numbers of soldiers, physicians or geographers, we'll know that they and we have failed.

Preparing for terrorism and its unavoidable consequences will require better tools for estimating casualties from blast effects and contamination of air, water and food supplies. GIS already has proven its worth in environmental risk modeling, often considering "worst-case" scenarios that pale by comparison to what happened on Sept. 11. Now those same tools must be applied to purposeful contamination of streams, reservoirs, wells and public utilities as well as purposeful disruption or contamination of the food chain from farmer to consumer.

For the GIS community as a whole, our greatest contribution will be to keep doing what we've been doing, especially in regard to infrastructure. We'll need to map and analyze communication networks, settlement forms, and utility and transportation networks to determine vulnerabilities and recommend changes. Ultimately, we'll need to predict troubling scenarios of disruptions and consequences and propose solutions. We'll need to consider new geographies of scarcity covering resources, goods (e.g., water, food, energy, etc.) and services.

We'll need to collaborate with economists and others to anticipate the economic stress caused by changes in resources, policies and behaviors. In the end, social change will be inevitable, and we'll need to collaborate with social scientists to help ease the transition. Already, concerns are being expressed for the loss of privacy and personal freedom that may follow. It will be imperative for geographers and the GIS community to "weigh in" on issues that involve our technology. Global Positioning System-based tracking of individuals is a prime example.

Public Awareness

Finally, it's important to anticipate what effect the attacks are having and will have on geography. Millions of viewers heard the term "GIS" for the first time when CARSI's maps and images appeared on television. A few days later, NBC's Katie Couric turned to Joseph Albaugh, director of the Federal Emergency Management Agency, and said, "So you're using something called GIS, or geographic information systems ..." That's a powerful introduction to the public, and it's bound to have an impact. Citizens of cities across the nation, for instance, may appreciate and support municipal GISs as never before.

Comedian Paul Rodriguez once observed, "Sometimes I think war is God's way of teaching us geography." Now add GIS to that dictum.

(This column originally appeared in the November 2001 issue of GeoWorld*)*

GIS Needs to Be Part of the Solution

By Jeremy Crampton

I received many letters supporting and challenging my thoughts on [homeland] security (see "Rethinking GIS and [Homeland] Security"). To elaborate, I argue that the GIS industry is in danger of becoming a "research arm of the security industry. Citing Monmonier (2002), I suggest that GIS is part of the problem as well as part of the solution.

There hasn't yet been a national debate about what kinds of freedoms we're in danger of giving up in the name of security. Even the last presidential election in 2004 did not put homeland security, technology and terrorism squarely into the public eye. Because of the subject's interest, I thought I would inform readers of a seminar I recently taught called "Maps, Politics and War."

War Stories

The seminar's focus was on the many interesting links between mapping and the pursuit of warfare. It began with some anthropological studies of war to see whether we are, as a species, naturally warlike (prone/programmed to fight?).

The seminar also was interested in war and politics, so we read about conflict in the Middle East and World War I. These case studies provided clear examples of how geographical borders and boundaries have real political outcomes. Which side of the line you're on (e.g., inside or outside the Gaza strip) can shape your whole life.

People treat you differently depending on where you're from. Many of the political geographers we read noted that geographical borders aren't so much "data" that are "captured" by maps and GISs, as GIS textbooks tell us. Rather, GIS is part of the process of constructing the social reality of these lines.

At the end of the seminar, we wanted to identify how geographers could use GIS to more positively understand and prevent terrorism, and we created the following list:

Contributions of Geography

1. Critical Analysis of Discourse

a. Critically analyze flows of knowledge—how "we" learn about and understand other cultures and vice versa, including the environment, culture, religion, language, customs, etc. Flows of knowledge, such as daily news, are never unconstrained by other factors.

b. Develop a historical perspective on these relationships and flows of information to understand how groups have been constructed by themselves and others (e.g., Arabs, "the West," "the Orient," etc.).

c. Acknowledge and "contextualize" the competing characterizations of terrorism (e.g., that it's to "overturn our freedoms" or is "religiously inspired" or is "endemic in the Middle East"). Terrorist experts define terrorism as being driven by the goal of political change, but often the national debate sees it as being driven by irrational hatreds. The political goals are usually quite specific and geographically contingent.

2. Origins of Terrorism

a. Acknowledge and analyze the many different origins of terrorism, especially and primarily its political origins and aspirations.

b. Document and understand the causes to show particularities and differences in terrorism with an emphasis on scales, types, and historical and present terrorist activities.

c. Relate these particularities to globalization and its effects, and analyze the multi-scalar causes and effects of terrorism.

3. Critically Informed Development of a Geospatial Database

a. A natural response to a problem such as terrorism is to amass as much information about it as possible, share the information among problem solvers and remove as many legal obstacles to information acquisition as possible. In this vein, many geographers have suggested that a geospatial database be constructed. However, many seminar members were hesitant to endorse spatial information gathering with such a "blank check."

b. Develop a non-partisan, critical voice to analyze the effects and repercussions of such a database on privacy and rights.

c. Promote awareness and responsibility regarding the creation of a geospatial database

4. "Geo-lobbyists"

a. Social scientists who depend on spatial information rarely take part in the political process of identifying, promoting and funding initiatives that would use their expertise. Therefore, we recommended that organizations such as the Urban and Regional Information System Association and the Association of American Geographers hire political activists in Washington, D.C., and government to represent the geographic component of terrorism.

b. We also wanted to see an increase in political activism among practicing geographers, GIS users and related fields.

This list was designed to help inform the public debate about the relationship of geography and terrorism. By opening our eyes to these relationships, perhaps GIS can become more of a solution rather than part of the problem.

(This column originally appeared in the July 2004 issue of GeoWorld.*)*

Experience the Power of PPGIS

By Nigel Waters

I recently fulfilled a long-term goal and attended my first Public Participation GIS (PPGIS) Conference. Sponsored by the Urban and Regional Information Systems Association (URISA), the Third Annual PPGIS Conference was held at the University of Wisconsin-Madison from July 18-20, 2004, in The Pyle Center overlooking beautiful Lake Mendota in downtown Madison, Wis.

The PPGIS conferences are small. The number of registered attendees was about 150—just a little more than 1 percent of the registrants at the annual ESRI gatherings. There was no conference theme, but the papers were divided into four tracks: Data, Organizational and Policy Issues; PPGIS in Rural Settings and Small Communities; PPGIS Theory, Science and Scientific Methods; and PPGIS Practice and Implementation.

Each track ran five concurrent sessions. In the Focus Group Meeting, several participants, myself included, felt the tracks were somewhat a "Procrustean bed" into which the papers and sessions had been awkwardly forced. It was suggested that a series of keywords identified by paper authors might help with the perennially difficult process of paper positioning and allocation.

Opening Session

The first session I attended was in the Science track, and the first paper in that session was presented by Lew Hopkins from the University of Illinois at Urbana-Champaign. Hopkins discussed the design and construction of new hardware that would allow PPGISers to gather around a large, horizontal screen to debate the pros and cons of new community plans. It was a truly innovative device that could enhance dialog and a sense of involvement.

The second paper described the mechanics of developing a 3-D GIS database to allow planning participants a better view of putative planning proposals. The session's final paper was an intriguing discussion of conflict management. This was a groundbreaking methodology, but use of the phrase "detecting destructives," when referring to dissenting voices, seemed unfortunate. To label the participants in communicative planning that don't agree with the consensus as "destructives" seems to be contradictory to what PPGIS is all about.

Keynote Address

Mel King, an emeritus professor from the Massachusetts Institute of Technology and executive director of the South End Technology Center, Boston, delivered the keynote address. His talk on “The Power of the Byte” noted that technology use for advocacy planning and community empowerment was nothing new.

PPGISers could learn much from those like King, who “ tilled this soil” decades before the term PPGIS first was used. Steve Ventura and Laxmi Ramasubramanian, conference co-chairs from the universities of Wisconsin-Madison and Illinois at Chicago, respectively, introduced King and noted that he had the privilege of first introducing Nelson Mandela to the American people.

More Sessions

Other paper sessions on the first day included talks on the “Interactive Community Evaluation of Surface Soil Contaminants” in the city of Buffalo, N.Y., and an “Early Warning System for Predicting Housing Abandonment” that raised troubling questions relating to privacy issues and data confidentiality. How would you feel about a Web site predicting that the houses on either side of your property likely were to be abandoned in the immediate future?

Then I moderated a session on PPGIS and Urban Design. The first paper on “Public Participation in Electrical Line Transmission Placing,” by Andrea Beck and Brian McGee of the American Transmission Co., Madison, seemed somewhat out of place. It was nevertheless a masterly demonstration of the technology’s true power, quintessential GIS.

Albert Benedict from the Center for Neighborhood Technology in Chicago then talked about using PPGIS to select transit-oriented development sites. I concluded the session with a paper, co-authored with Kathy Tang from Nipissing University in Ontario, Canada, that discussed the use of interactive Web-based PPGIS for soliciting feedback from citizens on a transportation plan in Calgary, Alberta, Canada.

Ramasubramanian and Aimee Quinn from the University of Illinois at Chicago delivered one of the last papers, “Participatory GIScience: Progress or Illusion.” This was provocative because it emphasized science in a field that might be expected to emphasize qualitative methods, and it underlined the need for evaluation methods to determine just how successful PPGIS is in practice.

Town Hall and Special Sessions

The conference wasn’t just about delivering academic papers and case studies. One of the more lively sessions was the Town Hall Meeting on opening day. Renée Sieber, McGill University, chaired a panel that sought to determine just how large the PPGIS “tent” should be. How should PPGIS be defined, what should be inside, and what could be left outside the tent?

Panelists included Mike Shiffer of the Chicago Transit Authority, Bill Huxhold of the University of Wisconsin, Angie Lee from ESRI and Mel King. The panel also considered how the various actors in the PPGIS universe—professors, community directors, vendors, librarians, state employees, etc.—might help each other do their jobs more effectively.

Focus groups from the four tracks met to draw together the various threads of the event’s discussions. The International Focus Group emphasized that PPGIS should be "demand driven." Technology shouldn't be the focus—PPGIS should be about the community.

The Rural Group's reporter emphasized the distinctions in rural areas, including different demographics, greater population homogeneity and, in most cases, less access to technology. The Science Group wanted, in part, to see greater reliance on qualitative methodologies, and the final group, charged with Practice and Implementation, bemoaned the move by state governments to increase the cost and difficulties of accessing local datasets, an ominous cloud on the PPGIS horizon.

Summarizing some of the different threads into an Action Plan for future conferences resulted in suggestions for more "how-to" workshops, manuals and a guide to "best practices in PPGIS."

Future PPGIS Conferences

Renée Sieber co-founded the PPGIS conferences that are a true gem in the annual round of GIS meetings. Attendees have a strong sense of commitment that GIS, along with involvement from "grass-roots" organizations, can make the world a better place.

If you have never organized a conference, then you don't understand all the work involved. I've organized several during my career, and each time I promise never to do it again. Fortunately, such an attitude isn't shared by Chris Dionne, URISA's director of conferences, and Pat Francis, URISA's conference and meetings coordinator.

You can visit the URISA Web site at *www.urisa.org* to order a CD-ROM of the proceedings from past conferences or learn about an upcoming conference. You also can join the discussion by visiting the open PPGIS forum at *ppgis.iapad.org*.

(This column originally appeared in the September 2004 issue of GeoWorld.*)*

How Can We Combat Application Erosion and Bloatware?

By Nigel Waters

American software tends to be like American cars: Small models are introduced as a cost-savings measure, but, as the years go by, the models tend to get larger. In software literature, this is called "creeping featurism" or "bloatware."

Some years ago, I complained to a former GIS graduate student about the problem, but he demurred. He couldn't wait for the latest version of his word-processing, spreadsheet or GIS software. He advocated the so-called "feature-keen" view, while I espoused a "feature-shy" position.

Concerns about bloatware include a steeper learning curve for neophytes, new interfaces for existing customers to learn and the recently recognized problem of application erosion, which can be defined as a steady decline in a system's used functionality.

Measuring Bloat

The Online Computing Dictionary defines software bloat as "The result of adding new features to a program ... to the point where the benefit of the new features is outweighed by the extra resources consumed (e.g., RAM, disk space or performance) and complexity of use." The large number of unused features contributes to subsequent application erosion.

Joanna McGrenere and her colleagues at the University of Toronto and the University of British Columbia conducted an exhaustive study of software bloat (*www.dgp.toronto.edu/~joanna/papers/gi_2000_bloat.pdf*). Their research used a mixed methodology of quantitative and qualitative procedures, studying 53 Microsoft Word and Office 97 users. They wanted to determine whether software users were overwhelmed by the number and variety of choices in the interface and whether they had coping strategies for handling such issues.

Bloat was assessed quantitatively by counting first-order functions (there were 265) and determining whether a significant portion was unused. By this criterion, the word-processing software was severely bloated. More than 15 percent of the functions weren't used at all, and only about 3 percent were used on a regular basis by more than three-quarters of the participants. Users were familiar with many more functions than they actually used. Such familiarity is likely to reduce their perception that the software is bloated, but it may be a sign of impending application erosion, and why pay for it if you don't use it?

McGrenere's research identified the feature-keen and feature-shy personalities previously mentioned. The former tended to be power users and the latter expert users. There's a subtle difference between the two groups. In the word-processing world, the former is characterized by secretaries and administrative assistants. Such individuals need and want the extra functionality, and, because they use the functionality all the time, they're unlikely to experience application erosion. Expert users are defined not by their expertise with the software package, but by their knowledge of computing and/or their application area (e.g., a computer scientist or a lawyer).

Expert users fell into two groups: those who wanted all the functionality, but would have preferred to be able to customize the interface for speedy work, and a second group who just wanted a simpler system. In the GIS world, a need for limited functionality may explain the popularity of less-expensive packages such as those offered by MapInfo Corp.

McGrenere's study concluded that there was objective bloat (functions unused by almost every user) and subjective bloat (functions used only by certain subsets of users) in which one person's bloat is another's favorite function. The writers suggest that part of the problem may be resolved by a better interface.

Adjusting Interfaces

There has been extensive written work on how interfaces may be designed to make software complexity more palatable. Jenny Jerrams-Smith pioneered research on an adaptive user interface for Unix operating systems in 1991. Jerrams-Smith's recent work on the topic may be found on her home page at *www.dis.port.ac.uk/~jerramsj/*.

A major research initiative on user interfaces, albeit with respect to information infrastructure as opposed to software, was the "every-citizen" interface project. A book that resulted from the project, *More Than Screen Deep* (www7.nationalacademies.org/cstb/pub_screendeep.html), sought to help develop a more citizen-friendly interface to the National Information Infrastructure (NII). The goal was to ensure access to the NII "regardless of age, physical ability, race/ethnicity, education, ability, cognitive style or economic level" to overcome "digital divides."

Adaptive and Adaptable Interfaces

In her more recent co-authored work (*www.dgp.toronto.edu/~joanna/papers/JMcGrenere_CHI_2002.pdf*) and in her just-defended doctoral dissertation (*www.dgp.toronto.edu/~joanna/papers/JMcGrenere_Thesis_DoubleSided.pdf*), McGrenere distinguishes between adaptive and adaptable interfaces. Adaptive interfaces use intelligence automatically to adjust the interface to better suit each user's needs. The adaptable interface allows users to customize the interface's look and feel so control remains with users.

McGrenere's doctoral work tested the effectiveness of a modern adaptive interface (Microsoft Word 2000) against a Personal Interface that she wrote using Visual Basic for Applications software. Microsoft Word 2000's menus adapt to users. Non-use of an option will cause it to disappear from the menus. Users customized McGrenere's Personal Interface via a simple "add-a-function" technique, and they could "toggle" between the personalized and Microsoft Word 2000 interface. Statistical and qualitative analysis suggested a preference for the Personal Interface, but the adaptive interface also had support.

McGrenere suggests a mixed-initiative interface in which users have the final control, but can switch to a user-modeled, adaptive interface if they choose. This conclusion is supported by the work of Eric Horvitz (1999) on the LookOut system in which agent-based modeling is used to examine incoming e-mail to suggest scheduling appointments based on sophisticated regression and Bayesian probability analysis of users' past behavior. The software makes recommendations like a "courteous butler" and disappears from the screen if users ignore its suggestions.

GIS Implications

Many GIS software companies have been issuing radically new software with completely redesigned interfaces. For established customers, having a toggle switch between old and new interfaces might be attractive. Even more interesting is the possibility of offering a competitor's interface. Thus ESRI might be able to lure away MapInfo users (or vice versa) by offering their rival's former customers a toggle switch that allows them to move between the two interfaces. ESRI's ArcGIS now comes in three flavors (ArcView, ArcEditor, ArcInfo)—all support essentially the same interface but with different levels of functionality.

Application erosion is a more difficult proposition. Lynne Wardell, president of the Haddon Group, suggested that the problem is better termed "environmental erosion," because its root cause is the failure of users' organizations to provide the necessary technical-support infrastructure to address the problem (*www.haddongroup.com/Article_Application_Erosion.pdf*). The problem can be alleviated by bringing GIS users together with support personnel from developers, training departments and technical support organizations. Blue Marble Geographics, however, sees this as an admission of failure (*www.bluemarblegeo.com*).

It also should be noted that Frank Linton and his colleagues have developed a recommender system that addresses erosion and new learning (*http://ifets.ieee.org/periodical/vol_1_2000/linton.html*). The system, "Organization-Wide Learning," examines and then recommends command usage to users. Although the systems have some privacy concerns, they represent an exciting, automated form of the old "buddy system" for learning new software.

(This column originally appeared in the October 2002 issue of GeoWorld.*)*

GIS and Criminal Shadows

By Nigel Waters

In the future GIS may be as much a part of the mainstream as desktop mapping packages, spreadsheets and even word processing packages. But it is not yet, and that was made obvious to me when I read David Canter's (1994) fascinating new book *Criminal Shadows*.

Canter is a psychologist with the University of Liverpool, England. He has published extensively in the area of environmental psychology, but his claim to popular fame is based on *Criminal Shadows* and his 10 years of successful research in creating psychological profiles of rapists and serial killers for various police forces in the United Kingdom. His work is similar to that carried out by the Behavioral Science Unit (BSU) at the FBI Academy in Quantico, Va. Canter has been a guest of the BSU, whose methods are made widely known through Jodie Foster's character, Clarice Starling, in the film and novel *The Silence of the Lambs* (Harris, 1988).

Canter's Use of Maps

One of Canter's more intriguing techniques is his use of maps. In *Criminal Shadows* he explains how he became involved in investigating criminal activity. He was sitting on a train in 1986 looking at a newspaper report listing a series of sexual assaults that had been committed in the previous four years in the London area. Canter first reorganized the data spatially as a map. Then he separated the crimes into individual maps—one for each year—and then he pointed to an area in North London and told the police where the offender lived.

Canter reinvented a mechanical GIS. The activities of the offender, John Duffy, showed both a temporal and spatial progression. The early offenses occurred within a small radius of his home. As the months passed he moved outward from his home base. The three murders were committed at the most distant locations. Canter's work is reminiscent of the classic studies touted as forerunners of GIS work, including Dr. John Snow's research in finding the source of a cholera outbreak in Victorian London. Canter's use of overlays is identical to the use of coverages or layers in a modern GIS. Canter acknowledges his debt to geographers who work on the spatial manifestation of criminal activity, including David Herbert, but in the book he makes no use of GIS.

GIS would prove an even more powerful tool than the maps Canter uses because it could include additional information that might reveal new patterns, and it would allow additional types of analysis. Finding the location of the criminal's home may be seen as a single-facility, location-allocation problem. That is the type of problem that now can be solved with the analytical routines in a GIS.

More recently, Canter (Brown, 1994) established that such offenders fall into several types, including "marauders" who have the shortest "home range" and "commuters" who offend farther away. The inclusion of transportation routes within a GIS being used for criminal investigation would be an obvious benefit for catching the second type of offender. Exploring the cognitive maps of offenders and exploiting past work in geography on mental mapping may prove fruitful for distinguishing the home range of one group of criminals from that of another.

The use of GIS to fight crime is becoming more and more widespread. For example, the police department in Calgary now uses MapInfo, GIS software from MapInfo Corp., Troy, N.Y. Other police departments also use the technology.

Canter's Use of Databases and Statistics

In his book, Canter (1994) relentlessly extolled the scientific nature of his methodology. Like his colleagues at the FBI, he developed what is known as a psychological profile. How is this done? Quite simply you create a data matrix or database with the rows as the known offenders and the columns as the characteristics defining the offense. You then use some statistical measure of similarity to put together known offenders who are most similar. Finally, you use a technique such as cluster analysis or multidimensional scaling (MDS) to put together groups of similar offenders. It is not that simple, because the measure of similarity can be difficult to choose as can the number of clusters or the dimensionality of the MDS space. But the methodology is well-established in the literature. You also can create Latent Descriptor Indices, as the FBI does, and then perhaps use a discriminant function to compare a new crime to known offenders.

Finally, you might use expert systems to computerize all this knowledge into if-then rules and predict the characteristics of your unknown offender—just don't give him or her the code. One might even use knowledge elicitation programs to elicit information from past offenders to build such an expert system (Waters, 1989).

Canter's unswerving support for the scientific method is moderated a little toward the end of his book when he debates the advantages of different models of psychology and appears to come down to support recent narrative approaches based on the "stories" or models that inform and guide our lives.

These approaches have been used by researchers in artificial intelligence, such as Roger Schank, to develop "scripts" for providing computers with an understanding of the world. Perhaps the last word should go to Ted Mellnik, a staff writer at The Charlotte Observer who uses GIS to map crimes in Charlotte, N.C. He told me, via E-mail, that reporters can write good narratives, but they still need a GIS to tell them where the story is.

Update: Canter wrote a follow-up book, *Mapping Murder: Walking in Killers' Footsteps* (Virgin Books, 2005).

(This column originally appeared in the January 1995 issue of GIS World.*)*

Peace on Earth: Using GIS for Military Purposes

By Nigel Waters

When I first began attending the big new GIS conferences, I noticed that the military and associated organizations were there in abundance. This particularly struck me at the 1988 GIS/LIS meeting in San Antonio, Texas, where there were a number of papers dealing with the applications and use of GIS for military purposes. Not all of the papers made it into the proceedings or were even available to the conference participants by request.

One such paper discussed the use of an expert system interface to a GIS to help locate land mines—an ideal GIS application. The GIS essentially stored the knowledge base, e.g., the shape of the terrain. I hope I'm not letting any secrets out, but the program suggested that a narrow pass was a good place to put a mine and that open plains were a bad place, etc. The expert system supplied the rule base—if narrow pass then land mine is true, etc. Obviously the rules could be more sophisticated, but I was left wondering after the presentation: What if the bad guys have the same system? I know, I know—a few random bumps in the locations should really fox them. As S. Brand (1992) notes: "The best funder of basic science in Washington is the Pentagon."

Alternative Roles Required?

Articles critical of the use of computers for military purposes appeared in a wide variety of places in the early 1980s, including journals such as *Datamation* and *New Scientist*. More recently, while much of the establishment has eulogized the use of computer and even GIS technology in the Gulf War (Schulman, 1991), others have been bitterly critical. "The war against Iraq in 1990-1991 was the first full-scale GIS war," says N. Smith (1992). "It put geography on the public agenda in a quite palpable if impalatable way as it claimed an estimated 200,000 Iraqi lives."

Smith notes later, "A considerable percentage of students who learn GIS in U.S. classrooms graduate to military and related jobs," and he states that the Defence Mapping Agency (now the National Geospatial-Intelligence Agency) is the largest employer of U.S. geography graduates. Laments Smith, "There is a curious lack of reflection on this military-geographic connection."

Given the use of GIS in the Gulf War, Smith finds S. Openshaw's (1991) appeal to GIS practitioners that they have “a moral duty to help society” as disingenuous. But, in fact, Openshaw is right.

Indeed, many writers, including Brand (1992), a former U.S. infantry lieutenant, have suggested alternative roles for the American military as part of the so-called “peace dividend.” Brand suggests that the U.S. Army might assume a strong role as a protector of the environment. Some of his examples, such as United Nations-sanctioned military intervention against the government of Burma (Myanmar) for its exploitation and destruction of the teak hardwood forests, might stretch our definitions of what constitutes internationally unacceptable behavior. But others, including humanitarian aid to Bangladesh typhoon victims and recent aid to those who lost their homes in the fury of Hurricane Andrew, seem to be universally meritorious. As with many natural disasters, those providing aid could benefit greatly from GIS use.

J.L. Petersen (1992) suggests various new roles for the U.S. military. His recommendations for international involvement may smack of a Pax Americana, but it is his call for the domestic involvement of the military in solving the miseries of the inner cities that is most interesting. He notes that in the past decade there have been more than 100 successful programs that involve the U.S. military in helping resolve urban problems. He details some of these programs, including projects in the 1990s—Save America’s Youth, Youth Service USA and Green Team. Many of the skills that the military provide are hard engineering skills, but also important are physical distribution, logistics, operations research and, of course, information systems and computer skills.

This is not new. There have been models in the past where organizations with a heavy military involvement have used their abilities in this fashion. Perhaps the best example is the work of the Rand Institute, which in 1968 began cooperating with New York City, and continued to provide engineering, logistical and information systems solutions for the next seven years.

During the 1970s, one of the great criticisms of the use of positivism in the social sciences (and many GIS applications fall into this camp) was that it lacked heart. Hence the rise of more socially aware approaches, including humanism and structuralism. There is no inherent reason, however, that the use of computers in general, and GIS in particular, should support the existing status quo. Many social justice groups, causes and organizations would benefit from the use of both aspatial and spatial databases. So if you are going to ask what you can do this Christmas with your GIS-related talents in your spare time, you might think about enlisting them in serving the common good.

CompuMentor (*www.compumentor.org*) based in San Francisco, is one organization that matches those with computer expertise with non-profit organizations needing computer-based help. Give CompuMentor a call to receive information on how you can do some good with your computer/GIS skills. The military already has a head start on you.

(This column originally appeared in the December 1992 issue of GIS World.*)*

GIS and Public Health: Are We There Yet?

By Jeremy Crampton

The conjunction of GIS and public health seems like an ideal match. Public health focuses on the spread and causes of diseases such as cancer, obesity and heart disease, which are among the nation's leading causes of death.

Public-health analysts accomplish this not by experimenting in labs, but by combing through socio-economic census data or looking at how social networks are linked to disease transmission. Such studies then can be used to make policy recommendations.

Obese Example

A public-health example is the nation's troubling increase in childhood obesity. Although obesity has a genetic factor, it's becoming increasingly clear that several environmental "risk factors" are chiefly to blame for rates of overweight children that have tripled since 1980.

Today about 9 million children are overweight, which shortens lives and helps increase diseases such as diabetes. The terms "obese" and "overweight" are defined differently for children and adults, but are based on body-mass index.

There's no single cause of childhood obesity, but geography seems to play a role. The Cartographic Modeling Laboratory at the University of Pennsylvania, for example, used handheld Global Positioning System technology to map food opportunities ("fat traps") near elementary schools in Philadelphia.

In an Associated Press (AP) story on this work, Amy Hillier, the lab's research director, was quoted as saying, "What we're doing is connecting people to the context in which they live." If kids have to navigate a neighborhood bursting with sugary snacks and sodas, they might be more likely to become obese.

This example shows how important geography is—not just in the sense of where things are, but also the characteristics of the neighborhoods and the places where people live. Rural childhood obesity is increasing faster than urban obesity, for example, perhaps because high-calorie diets from several generations ago, when people physically worked the land, still are being handed down to people increasingly accustomed to mechanized agriculture.

Potential Problems

Obviously, mapping and GIS should play a major role in public-health research. But a recent study shows that public-health analyses aren't using the potential of mapping and GIS to the fullest. Stacey Martin, a geographer at Georgia State University, analyzed the types of maps used in six public-health journals and one cartography journal. Martin looked at all the articles in the journals published between 2000 and 2004, and her findings make interesting reading (Martin, 2005).

First, the good news: across all journals, the number of maps has increased tremendously from about 50 per year in 2000 to about 250 per year by 2004 (a five-fold increase). In all, about 875 maps were published. It's not clear how much of this increase is due to a rising awareness of mapping to present and analyze findings, but it surely represents encouraging news for GIS.

Comparing the six public-health journals to the one cartography journal, more maps were published in the cartography journal than in any single public-health journal, but this is to be expected. In fact, the cartography journal was included as a deliberate comparison to see how well the public-health journals "stack up" against a professional mapping journal.

It turns out that the public-health journals accounted for more than 60 percent of all the maps published, but that the number of maps varied from journal to journal. One public-health journal published as few as 11 maps, while another published 186.

Map Types

Next, Martin looked at the types of maps that were published, and the results were disturbing. Martin established 11 categories of maps and looked at how often the map types were used among journals (see Figure 1).

The single most popular type of map used by the public-health journals was the choropleth map. No other type of statistical map came remotely close, and, among public-health journals, this one map type accounted for between one quarter and three fifths of all maps published. The cartography journal, although it published twice as many maps as the most "map-happy" public-health journal, used the choropleth a scant 13 percent of the time. I believe the choropleth map is one of

JOURNAL	AEP	AJPH	AJE	PHGIS	IJHG	HP	CaGIS	TOTAL
TOTAL MAPS	11	42	45	113	186	136	342	875
Choropleth	4	17	11	46	113	64	44	299
Reference/Location	1	16	16	34	5	44	59	175
Base Map	0	0	0	9	51	0	33	93
Raster or with raster-like elements	2	0	8	7	2	12	62	93
Geo-technique	0	0	0	0	0	5	60	65
Isarithmic	0	0	7	11	4	3	22	47
Projections	0	0	0	0	0	0	38	38
Dot map	3	2	0	1	8	0	8	22
Proportional symbols	1	4	3	2	2	8	2	22
Dasymetric	0	0	0	1	1	0	15	17
Combination - Choro/Dot	0	3	0	2	0	0	0	5

Figure 1: A study looked at how often map types were used among journals. (MARTIN, 2005)

the worst map types with which to present data. Martin's study shows that it's still surprisingly popular outside of cartography and GIS.

Figure 1 also shows that the "cartodiversity" of map types is dangerously low among public-health journals. Although the cartography journal demonstrates a healthy range of mapping types, public-health journals are either unaware of or don't understand the full range of mapping possibilities.

Not one public-health journal published a map projection, for example, but any introductory cartography book will correctly note that awareness of map projections is essential for correctly understanding and visualizing spatial data.

Public-health professionals are good at educating the public about dangers to their health. What's needed now is a similar educational campaign for public-health professionals themselves in the use of mapping and geovisualization methods.

(This column originally appeared in the July 2005 issue of GeoWorld.*)*

Why We Shouldn't Collect or Use Race Data

By Jeremy Crampton

A French law passed in 1978 explicitly bans the collection or computerized storage of race-based data without the consent of individuals or a state-based waiver. Imagine if the same law was in the United States—no GIS analysis using racial variables, not even collecting race-based data in the U.S. Census.

Why would the French pass such a law? Surely race-based data are needed to eradicate social inequalities, direct public-health policy and ensure civil-rights compliance? Don't we need such data to perform GIS analysis?

In fact, there are reasonable arguments for supposing that none of this is true.

The U.S. Census

Let's start with race-based data collection. The U.S. Census has two questions relating to ethnicity and race:

1. Is Person 1 Spanish/Hispanic/Latino?
2. What is Person 1's race?

What is race? Many people today view race as natural and separate divisions within the human species. This view is reflected in the Census form where people are asked to identify their race in the following Census Bureau categories:

- White
- Black, African American or Negro
- American Indian or Alaska Native
- Six Asian categories, including Chinese and Asian Indian
- Three Pacific Island categories

Although, for the first time, the 2000 Census allowed people to identify themselves as belonging to more than one race, less than 2 percent of Americans did so. So about 98 percent of Americans see themselves as belonging unambiguously to one racial category.

The Anthropological View

Anthropologists, however, have stated on numerous occasions since World War II that race doesn't exist. Consider the following official statement from the American Association of Physical Anthropologists:

"Pure races, in the sense of genetically homogenous populations, do not exist in the human species today, nor is there any evidence that they have ever existed in the past ... Humanity cannot be classified into discrete geographic categories with absolute boundaries" (*www.physanth.org/positions/race.html*).

But if race isn't real, but rather an arbitrary construct not supported by geographical or genetic data, perhaps we should stop collecting race-based data? Furthermore, it's argued, by repetitively iterating these social categories, we're in danger of perpetuating them.

But perhaps the most relevant question in GIS is whether race is an explanatory variable at all—does it *explain* anything in and of itself, or should the focus be elsewhere? Attempts have occasionally been made in the United States to ban the collection of race data (most recently in California). They've done so in France, which instead focuses on class data (e.g., wealth, poverty, income).

And it's this point that affects GIS the most. In any country concerned with alleviating poverty and social inequalities, such as the United States, isn't it *poverty* that's the variable of concern? Isn't poverty the variable we should be addressing? Why throw "race" into analyses? Race isn't real the way poverty is.

Misconceptions and Counterarguments

First, let me note that race is a complex issue, and I'm not an anthropologist. But in this emerging public debate, we need to clear up some misconceptions. The following are some that I've come across:

1. "Races exist."

Race is a social construct, with no biological component, now or in the past.

2. "But people do vary."

Yes, humans vary from one another, but they do so on a biological and geographical continuum (cartographically analogous to an isoline map such as temperature). Cutting up the continuum into a small number of separate categories is easy enough and, indeed, has been done in many different ways, but it doesn't emerge naturally from the data—it's imposed.

3. "Don't the data support rough groups?"

In 1972, anthropologist Richard Lewontin discovered that genetic variation occurred largely within a population, and not between them. About 94 percent of genetic differences in human variation occur *within* populations. And group genetic differences aren't increasing, they're decreasing with continued human migration, so individuals today are mostly of mixed ancestry.

4. "What about race-related diseases such as Tay-Sachs (Ashkenazi Jews) and sickle-cell anemia (West Africans)?"

This argument is heard quite often (in GIS and public-health work, for example). Race data maps are needed, it's claimed, to identify areas where certain diseases occur.

This argument suffers from the weakness, however, that disease doesn't map on to racial groups, largely for the aforementioned reason: people today are of mixed ancestry, and ancestry isn't race.

Sickle cell anemia is higher in West Africans, but also in white Mediterranean peoples and even in some Asian-Indians. And not all West Africans are at risk. A public-health policy that targets African Americans and sickle cell anemia will therefore miss whites and Indians at risk as well as over inflate the risk (with the attendant worry and stress) to African-Americans.

Anthropologists and sociologists have professional statements on the use of race or race data. The latter note its contentiousness, but strongly favor retaining race data collection.

I recently asked my professional organization, the Association of American Geographers (AAG), whether it has contemplated a similar statement, perhaps preceded by an informed debate. It appears that they haven't. Perhaps the time has come for AAG and GIS organizations to address this issue.

(This column originally appeared in the April 2006 issue of GeoWorld.*)*

Section 4.
References

Arthur, W.B., 1994. *Increasing Returns and Path Dependence in the Economy*. University of Michigan Press, Ann Arbor, Mich.

Brand, S. 1992. "Army Green." *Whole Earth Review*, No. 76, Fall 1992, pp. 58-59.

Brown, J. 1994. "Electronic Detective Throws Ring Round Rapists." *New Scientist*, Vol. 143, p. 6.

Canter, D. 1994. *Criminal Shadows: Inside the Mind of the Serial Killer*. Harper Collins, London.

Chang, K. 2001. "From 5,000 Feet Up, Mapping Terrain for Ground Zero Workers," *The New York Times*, Sept. 23, 2001.

Cutter, S.L., Richardson, D.B., Wilbanks, T.J. (eds.). 2003. *The Geographical Dimensions of Terrorism*. Routledge, New York.

Demangeon, M.A. 1930. *L'Habitat Rural en Égypte*. Imprimé par l'Imprimerie de l'Institut Français d'Archéologie Orientale du Caire pour La Société Royale de Géographie d'Égypte, Paris.

Demangeon, M.A., Weiler, A. 1937. *Les daisons des homesde la hutte au gratte-ciel*. Éditions Bourrelier & Cie, Paris.

Dobson, J.E. 2003. "Estimating Populations at Risk," *Geographical Dimensions of Terrorism*, Chapter 5.5, Cutter, S., Richardson, D., Wilbanks, T.(eds.) Routledge, New York.

Dobson, J.E. 1979. "A Regional Screening Procedure for Land Use Suitability Analysis." *The Geographical Review,* Vol. 69, No. 2, pp. 224-234.

Dobson, J.E., Bright, E.A., Coleman, P.R., Bhaduri, B.L. 2003. "LandScan2000: A New Global Population Geography," *Remotely-Sensed Cities*, Chapter 15, Mesev, V. (ed.), pp. 267-279. Taylor & Francis, London.

Dobson, J.E., Bright, E.A., Coleman, P.R., Durfee, R.C., Worley, B.A. 2000. "LandScan: A Global Population Database for Estimating Populations at Risk," *Photogrammetric Engineering & Remote Sensing*, Vol. 66, No. 7, pp. 849-857.

Dobson, J.E., Trimble, S.W., Dobson, J.R., Friend, D.A. 1999. "In Memoriam: Kirk H. Stone, 1914-1997," *Annals of the Association of American Geographers*, Vol. 89, No. 3, pp. 535-548.

Dunlap, D.W. 2001. "Hey, New York. Nice Map. Care to Give It a Try? The City in

Exquisite Detail, Soon to Be Online," *The New York Times*, Feb. 15, 2001.

Ferguson, C.H., 2005. "What's Next for Google?" *Technology Review*, Available online at: *www.technologyreview.com/articles/05/01/issue/ferguson0105.asp?p=1.*

Harris T. 1988. *The Silence of the Lambs*, St. Martin's Press, New York.

Horvitz, E. 1999. "Principles of Mixed-Initiative User Interfaces." *Proceedings of the Computer and Human Interaction 1999 Conference*, p. 159-166.

Martin, S. 2005. "Cartography, Discourse, and Disease: How Maps Shape Scientific Knowledge About Disease," Master of Arts Thesis, Georgia State University. Available online at: etd.gsu.edu/theses/available/etd-04222005-094353.

Monmonier, M. 2002. *Spying with Maps—Surveillance Technologies and the Future of Privacy*. University of Chicago Press, Chicago.

Openshaw, S. 1991. "A View on the GIS Crisis in Geography, or, Using GIS to Put Humpty-Dumpty Back Together Again." *Environment and Planning*, Vol. A, No. 23, pp. 621-628.

Petersen, J.L. 1992. "Will the Military Miss the Market." *Whole Earth Review*, No. 76, pp. 62-70.

Rapoport, A. 1969. *House Form and Culture*. Prentice-Hall, Englewood Cliffs, NJ.

Schulman, R. 1991. "Portable GIS from the Sands of Desert Storm to the Forests of California." *GeoInfo Systems*, Vol. 1, No. 8, pp. 24-33.

Sieber, R. 2003. "Public Participation GIS Bibliography," available online at *http://deathstar.rutgers.edu/ppgis/*. (Accessed on March 19, 2006).

Smith, N. 1992. "History and Philosophy of Geography: Real Wars, Theory Wars," *Progress in Human Geography*, Vol. 16, No. 2, pp. 257-271.

Walker, H.J., Richardson, M.E. 1994. "In Memoriam: Fred Bowerman Kniffen, 1900-1993" *Annals of the Association of American Geographers*, Vol. 84, No. 4, pp. 732-743.

Waters, N.M. 1989. "Expert Systems Within a GIS: Knowledge Acquisition for Decision Support Systems." *Proceedings, Canadian National GIS Conference*, Ottawa, Canadian Institute of Survey and Mapping, pp. 740-759.

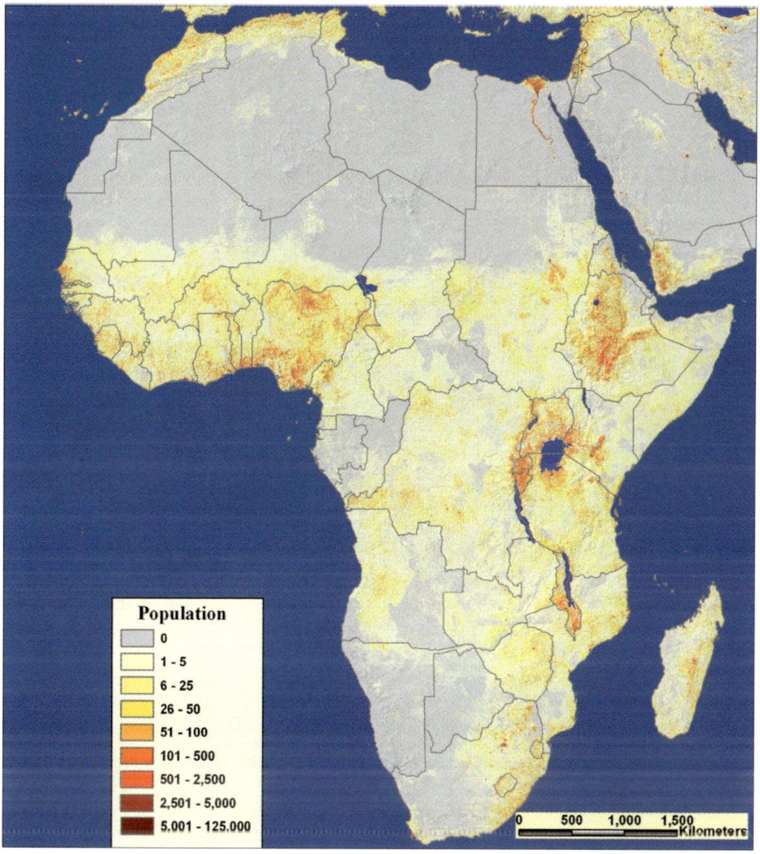

PLATE 1.
This LandScan population map of Africa shows the population density per square kilometer for this continent (see pg. 51).

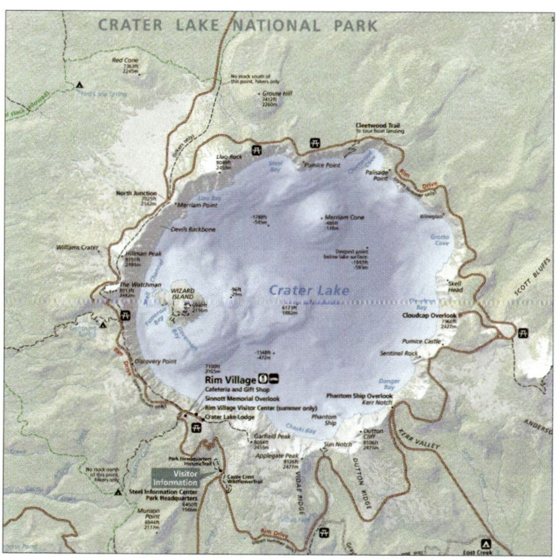

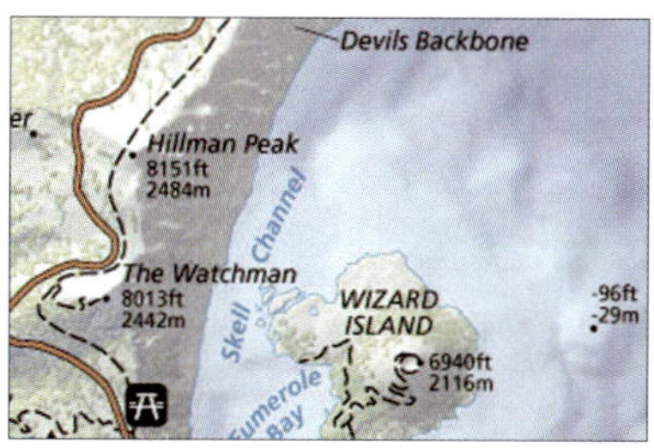

PLATE 2.
The National Park Service's Tom Patterson used modern technology and artistic methods to create a map of Oregon's Crater Lake (left). A detail of the same map (above) shows that by standing at "The Watchman," viewers could look down slightly (326 meters) across the Fumerole Bay and see those who followed the twists of the track to the peak of Wizard Island (see pg. 52).

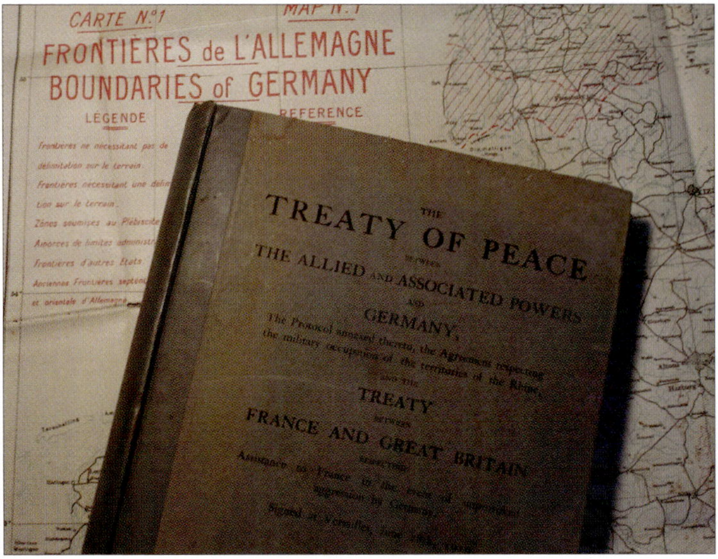

J.W. CRAMPTON

PLATE 3.
The Treaty of Paris signed at Versailles in June 1919 is shown lying on one of the official boundary maps delimiting the new territory of Germany (see pg. 59).

PROCEEDINGS OF THE NATIONAL ACADEMY OF SCIENCES

PLATE 4.
A cartogram, produced by Michael T. Gastner and M.E.J. Newman, distorts the United States based on population density (see pg. 110).

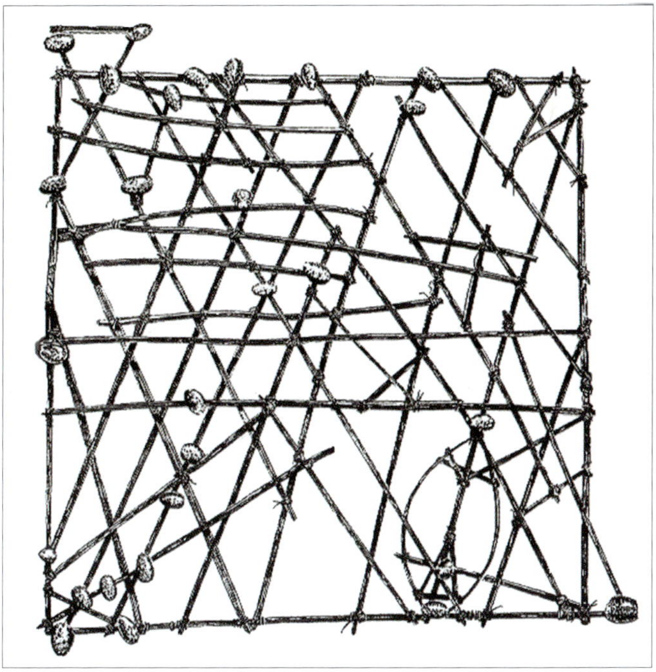

THE SMITHSONIAN'S NATIONAL MUSEUM OF NATURAL HISTORY

PLATE 5.
Marshall Islanders used three distinct types of stick charts to show the novice fundamentals about navigation, to show swell patterns in relation to a number of islands or a group of atolls and to show the overall spatial configurations of many islands (see pg. 116).

PLATE 6.
Geospatial technology has profoundly changed warfare with GPS-enabled precision bombing, unmanned aerial vehicles with high-resolution cameras (above), and troop tracking systems (left) (see pg. 127).

PLATE 7.
Commercial remote sensing satellites provide a valuable input into geospatial systems, yet they also seem to raise privacy fears. Common misconceptions about these satellites are that they are always on, that they can track moving objects and that the resolution is good enough to see fine detail such as facial features (see pg. 159).

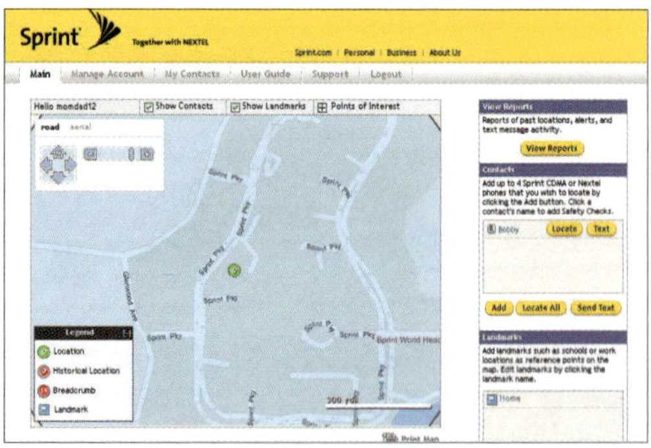

PLATE 8.
Tracking products and services, such as the Sprint Family Locator Service, are designed to provide parents and caregivers of Alzheimer's Disease patients peace of mind, but there's also the potential for misuse. Pictured here is both a phone and Web-based map interface that interact with the GPS-enabled phone that is carried by the child or patient (see pg. 167).

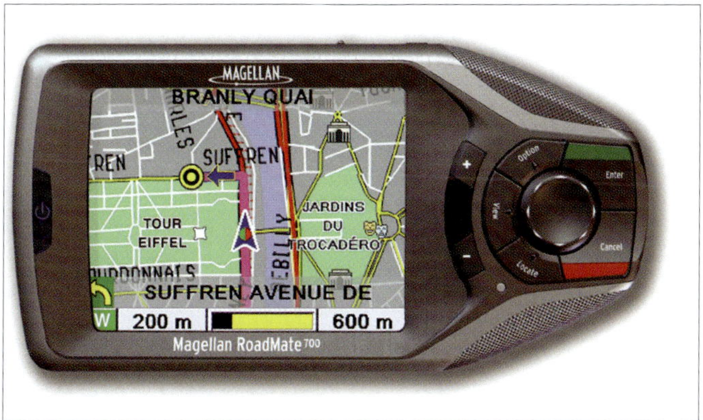

PLATE 9.
The replacement of paper road maps with onboard navigation systems for cars, such as this Magellan Roadmate 700, can be seen by some as an example of recomplicating a previously simple and straightforward task (see pg. 182).

PLATE 10.
Mobile mapping devices with integrated GPS come in many shapes and sizes. Handheld units such at the TrimbleRecon (above right) offer easy portability. Pen-based mobile options such as the Xplore Technologies' iX104 (above left) slate and Panasonic's Toughbook 18 (left) convertible tablet offer easy rugged options for bringing GIS to the field (see pg. 201).

PLATE 11.
Geographic analysis–at any scale, resolution or coverage–requires fieldwork, and GIS is no exception. Here an image of Sarajevo, Bosnia illustrates the dangers of populations at risk due to regional conflict (see pg. 215).

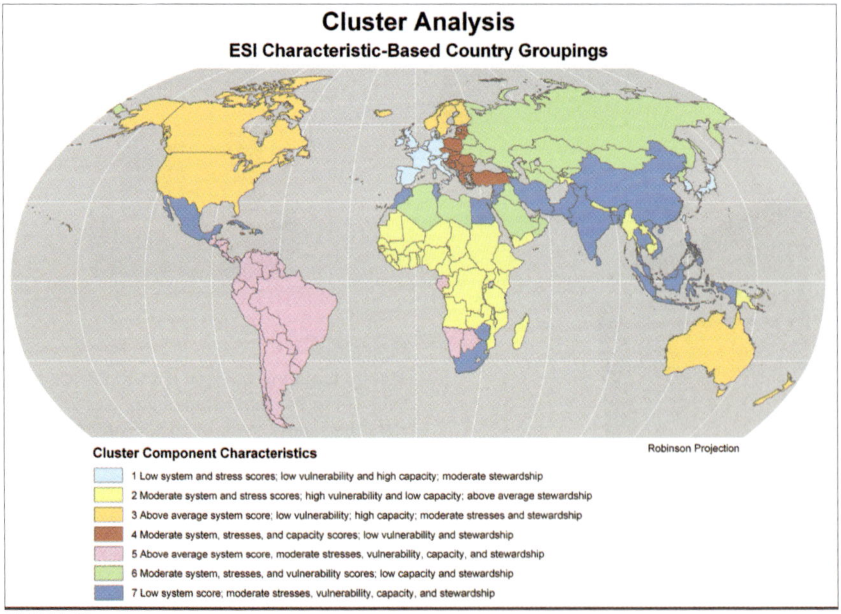

CREDIT: ESTY, DANIEL C., MARC LEVY, TANJA SREBOTNJAK, AND ALEXANDER DE SHERBININ 2005. 2005 ENVIRONMENTAL SUSTAINABILITY INDEX: BENCHMARKING NATIONAL ENVIRONMENTAL STEWARDSHIP. YALE CENTER FOR ENVIRONMENTAL LAW & POLICY, NEW HAVEN.

PLATE 12.
The Environmental Sustainability Index benchmarks the ability of nations to protect the environment over the next several decades (see pg. 217). (http://sedac.ciesin.columbia.edu/es/esi/)

PLATE 13.
Ubiquitous computing will enable diverse wireless applications, including monitoring of pets and houseplants, operation of appliances, keeping track of books and bicycles, and much more (see pg. 218).

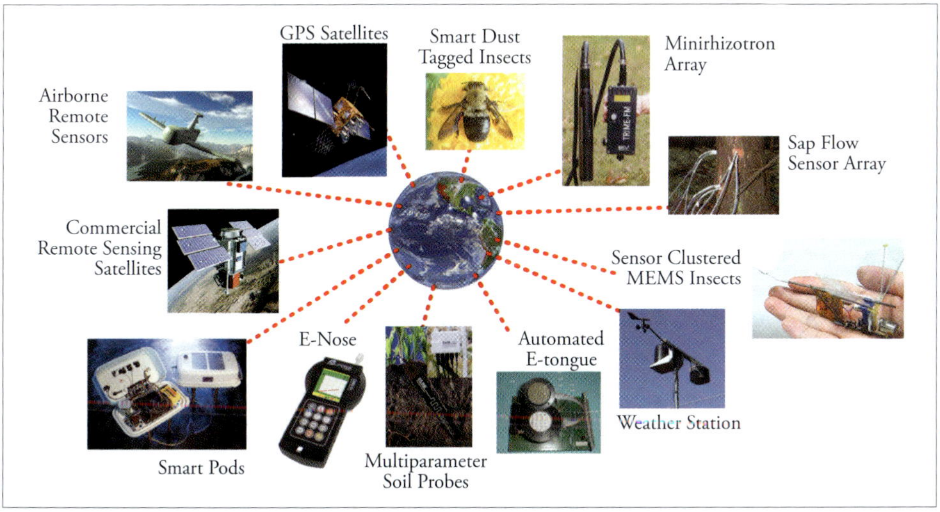

VINCENT TAO

PLATE 14.
With the presence of cheaper, miniature and smart sensors; abundant fast and ubiquitous computing devices; wireless and mobile communication networks; and autonomous and intelligent software agents, the Sensor Web has become a clear technological trend in geospatial data collection, fusion and distribution. The Sensor Web is a Web-centric, open, interconnected, intelligent and dynamic network of sensors that presents a new vision for how we deploy sensors, collect data, and fuse and distribute information (see pg. 24).

PLATE 15.
The idea of virtual reality has evolved over time from the headset and glove interface of the early 90s to a more accessible immersive Web experience. Sim City 4 from Electronic Arts is an example of a rich and realistic gaming environment. It includes over 100GB of topographical information from the United States Geological Survey and a wealth of real-life buildings that can be imported into this environment (see pg. 191).

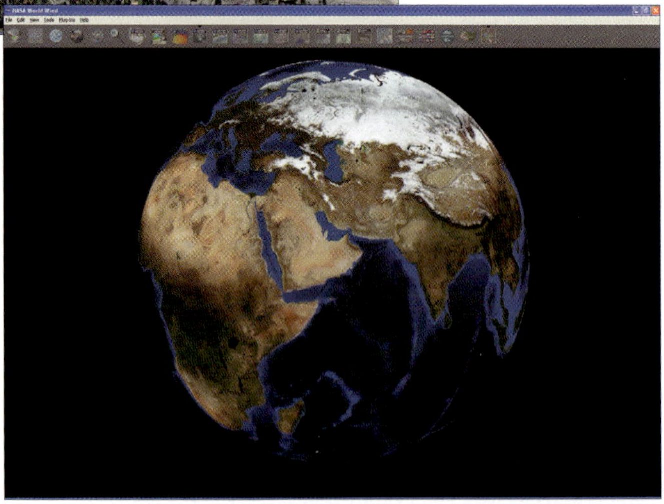

NASA

PLATE 16.
Geographic Exploration Systems, such as NASA's World Wind pictured here, afford the users a simplified and compelling interface to explore geospatial data. The advent of these browsers has resuscitated the idea of a Digital Earth that enables users to explore and interact with vast amounts of natural and cultural information gathered about the Earth (see pg. 233).

Section 5.
GIS, Ethics, and Education

Introduction
By Matt Ball

With great power comes great responsibility. The columns in this section remind geospatial technology practitioners of the purpose of these tools as well as to be mindful of their power.

Geospatial technology has a significant role to play in such noble pursuits as safeguarding public health, combating poverty, fighting crime and terrorism, and mitigating manmade and natural disasters. However, the technology is "surveillant" in nature (Waters, 1994), and safeguards are warranted to ensure that practitioners don't manipulate the power of the tools to harm others.

In Dobson's opening column in this section, he relates the incendiary nature of the privacy debate through the reaction to his column, "Consider Both Sides of GIS Ethics." That column drew the most responses that he has experienced and, in fact, some hate mail. The debate over privacy often trumps the social benefits of the technology, yet debate is critical to ward off acquiescence.

Waters (2001) speaks to the ethical applications of these tools for land-use planning issues such as environmental justice, the protection of land and resource rights, and collaboration with stakeholders through public-participation GIS. In these examples, the tools are harnessed to combat social injustice. In contrast, the ethics of practitioners lies in their responsibility (Obermeyer, 1998) to the ideals of the profession, their clients, society, and between the profession and a student.

Dobson warns of the dangers of human-tracking devices, coining the term "geoslavery" to describe the potential misuse. The threat of malicious use is very real, and cases of such abuse have come before judges in many jurisdictions. It's hard to imagine safeguards that wouldn't impede normal use. Sui's discussion of the "Laws of Location-Based Services" provides a moral compass to gauge when the line has been crossed.

Collectively, these columns provide a valuable awareness of the "dark side" of misused technology as well as the pitfalls of relying too much on a system when bad data, methodology or removal from the physical reality of a location may cloud decisions.

This section also provides a valuable perspective on the use of GIS in education. The benefits and potential downfalls of an earlier introduction of GIS tools in the classroom are discussed in Waters' column (1997), "Should GIS Stay Out of Our Schools?" The possible harnessing of GIS to aid in the education of boys is a particularly timely subject as the decline of boy's grade-school performance is a topic of strong debate.

Is GIS a Privacy Threat?

By Jerry Dobson

Should people be concerned about GIS as a privacy threat? The answer, according to GIS proponents and opponents, is a resounding "Yes!" The difference lies in how each side values societal benefits vs. costs. Most people admit there are mixed blessings, but they differ greatly in which aspects they emphasize.

GIS awareness is growing rapidly in the public consciousness, and privacy looms large as a debate topic. Indeed, GIS ethics often attract more attention among the general public than GIS benefits. To illustrate, since 1991 I've written 45 columns for *GeoWorld (formerly GIS WORLD)*. Only one of them focused on GIS ethics (Dobson, 1993), and that one attracted more public attention than anything else I've written in my entire career. It's quoted in the *Millennium Whole Earth Catalog* (Rheingold, 1995), was discussed recently on talk radio and triggered the only hate mail I received in eight years of writing for *GeoWorld*.

Someone to Watch over You

On March 14, 1998, Charlie Zdravesky of radio station KUNM 89.9 FM in Albuquerque, N.M., hosted an hour-long discussion of GIS. At least half the show addressed privacy issues, and Zdravesky's own bias came across in his theme song, "Someone to Watch over You," and his running commentary. He quoted my negative comments concerning privacy while omitting positive comments about GIS as a democratizing force.

Zdravesky interviewed 12 GIS practitioners from New Mexico plus eight outsiders prominent in national and international GIS affairs. Without exception, they expressed serious privacy concerns. Remarkably, the statements made by eminent GIS proponents were virtually identical to those by outspoken GIS critics. In addition, precisely the same examples were offered by both sides; the most common complaint was that geocoordinates and financial transactions are recorded simultaneously every time an individual uses a credit card. Interviewees noted that such information is used for targeted marketing, but no one suggested the more ominous uses that an alarmist may imagine for such detailed documentation of an individual's movements, actions and preferences.

Several interviewees stated that GIS' power lies in its ability to integrate diverse information from multiple sources. Thus, the perceived privacy threat results not from new information collected by GIS, but from new inferences that can be derived by associating geographic information with personal information. One example is the ability to link credit card transactions with public records (e.g., deeds, tax records, actuarial records, court documents), census data and other geographic and personal information to characterize the person who made a purchase as well as his neighbors.

Thus, as Mike Goodchild of the University of California at Santa Barbara said, you may receive a catalog because a neighbor purchased an item from that company last year. An even greater concern expressed by Tom Poiker of Simon Frazier University, Burnaby, British Columbia, Canada, is that individuals' health insurance rates could rise because they live in a neighborhood with a high incidence of AIDS, cancer or crime. Zdravesky pointed out the conflict between two fundamental values: privacy and the public's right to know.

Tradeoffs

I was among those who raised serious concerns about privacy, balanced with a pro-GIS stance on societal benefits. I said, "If you look back at the *1984* (Orwell, 1949) concept that Big Brother was going to take over, there were a lot of people concerned with computers and communication technologies. It all depends on how you view the control. If you think that some evil group or some inordinately powerful group is going to control it, then you have a reason to fear it." On the phone I added, "But I see GIS as a democratizing force, sharing that control with the people," which Zdravesky omitted from the broadcast.

When Zdravesky called to interview me, his principal worry seemed to be satellites. He mentioned the potential for monitoring southwestern farmers' water usage. The fear of satellite surveillance isn't uncommon, but it seems misplaced. I would argue that the benefits are many and risks are few from sensors that see only from above.

Gar Clark, GIS manager for the city of Santa Fe, N.M., voiced a common misconception. "We've got satellite imagery up there that can see the mole on my right thigh," he said. For a variety of reasons (i.e., spatial resolution, look angle, pants, etc.) even the finest of today's overhead sensors can't really find such features. If sinister forces indeed require an image of Clark's mole (an unattractive prospect at best), they'll find it faster, better and cheaper to stalk the speckled appendage with a hand-held camera.

Greater Threats Abound

Neil Smith of Rutgers University, New Brunswick, N.J., noted that "social surveillance isn't particular to GIS." In fact, GIS pales by comparison to many direct surveillance forms that are embraced by society. Consider the irony of fretting about satellites hundreds of miles above, yet accepting video cameras a few feet away in school buses, city streets, police cars and businesses. This surveillance has been primarily in public or business settings, but video cameras have reached as far as residential hallways in dormitories and apartment houses. Society accepts these intrusions because of perceived benefits, mostly addressing physical safety and material security, and there has been astoundingly little debate regarding social implications.

An obvious lesson is that we can't assume the public will resist. For a host of perfectly good reasons, we willingly accept surveillance in the most private parts of our lives. Recently, for example, when automatic flushing devices were installed in restrooms, the public accepted that form of very personal surveillance without reservation, much less dissent. Typically, no one asked the basic questions: What kind of sensor is employed? What's its spatial resolution? Are images retained? If so, how long?

The real threat is a creeping acquiescence to all sorts of intrusions, and debate is crucial. If, for instance, a radical minority of parents cried "Big Brother!" when the first video cameras appeared on school buses, society may have considered the profound implication of an entire generation growing up accustomed to constant surveillance. For people inclined to dismiss this implication, consider a currently popular Web site in which a young woman invites near total surveillance. A camera in her bedroom broadcasts digital images every two minutes whether she's present or absent, dressed or undressed, sleeping or making love. Yet there's no suggestion of prurience; her motivation seems to be a deeply held belief that every life is worth watching.

Of course, readers who search for the aforementioned site should keep in mind that their employers may monitor their Internet use, may object to certain sites and may terminate employment based on abuse.

Which is the greater surveillance?

(This column originally appeared in the July 1998 issue of GeoWorld.)

Consider Both Sides of GIS Ethics

By Jerry Dobson

Children are killing children. Doomsayers have warned us about wrathful gods, menacing armies, plagues, overpopulation, pornography, nuclear holocaust, and even a few vices I personally enjoy. (Once, when my wife asked for help deciding a food theme for a party, I suggested cholesterol. We figured not many people would come, but the ones who did would be a lot of fun as long as they lasted.) Yet, I do not recall anyone ever warning us about children killing children. I tend to read more on the optimistic side, so please forgive me if I missed that message somewhere between Jericho and Waco. [Update: It later occurred to me that William Golding had warned us in *Lord of the Flies*.] To me it is the most chilling fact, the most frightening vital sign, in an age with more than enough contenders.

Public opinion has settled on television and movie violence as the principal villain, but I doubt we will see a statistical nosedive following the recent decision to put warning labels on television programs. Undoubtedly, there are more important factors, such as breakdowns in family and community structures, but I feel compelled to mention a suspicion that relates directly to what you and I are doing with GIS.

In a typical video game a child kills 40,000 or even 100,000 targets. It doesn't matter whether the targets are bugs, spacecraft, monsters, or pedestrians. In most games, there is no ethic or emotion on either side of the exchange, not even the customary good guys/bad guys storyline of a movie, television program, or comic book. Unlike these traditional forms of entertainment, video games involve children as participants. The child doesn't know who the targets are or why they should be killed, other than to run up the score.

Think of this next time you watch reporters interview a child who has killed another child or a teenager caught for a drive-by shooting. These are not crimes of passion or greed. They are emotionless acts that seldom do more than run up the score.

Video games are not all that different from GIS. Both simulate spatial environments and impart complete control to an operator. Both offer rich representations of physical phenomena (real or imagined) and poor representations of cultural phenomena (especially ethics). Yet, GIS is a powerful factor in real-world events from local to global.

Ethics in GIS

The strongest charge yet made against GIS is Neil Smith's (1992) accusation that

GIS claimed 200,000 Iraqi lives in the Gulf War. I argue precisely the opposite. The Iraqi people entered that war with virtually no information. If they had understood world consensus and resolve, if they had known the numbers and technological advantages of forces massed against them, they might have recognized an external threat greater than the internal threat posed by their leader. Much of any dictator's power lies in control of information. Saddam Hussein survived the combined military might of the world's great powers, but I doubt he could survive a GIS in every Iraqi village.

To those who suggest that GIS cannot represent culture, I counter that the current deficiency is related more to society's priorities than to technical limits. Government and industry invest far more in analysis of infrastructure and physical resources than in cultural research. GIS suffers from this same bias due to its cost. GIS developers are making headway in certain cultural areas where funding is available, notably in market research. Expect more progress when we finally link GIS with process models, but development of physical models will continue to outpace development of cultural models.

Some people view GIS as a potential evil and fear centralized control. I view GIS as a potent, democratizing force. I remind detractors that the single greatest factor credited with the fall of communism in the Soviet Union, and almost in China, was information flow. This was suggested by many observers, and no one singled out a particular type of information. The implication is that more information leads to better societal decision making and broader consensus. If fax machines can produce a bloodless revolution of the magnitude witnessed in Russia, imagine what GIS can do.

A Risk Worth Watching

I used to think these defenses were sufficient, but video games and drive-by shootings have made me wonder. Is it possible that an information technology can transfer its values, or lack of values, to the operator? Remember this is just a suspicion. When I watch the shooters being interviewed, I don't know who plays video games and who does not, but I hope someone will study the phenomenon. [Update: Six years later, the shootings at Columbine High prompted many studies of the relationship between video violence and aggressive behavior. It appears that video games do encourage aggressive behavior in certain individuals. Overall, however, youth violence has decreased since video games were introduced.] If a connection exists, the GIS community will need to pay closer attention to ethics. The risk can only increase as we move toward virtual reality if we don't include ethical reality.

In retrospect, my very first GIS work (Dobson, 1979) taught me a lesson in environmental ethics. We developed an approach for siting noxious facilities, intended to resolve conflicts by enabling each interest group to formulate criteria based on its own values. Environmental interests were well represented. We applied the criteria to real data and found sites that optimized across all values. We felt good about the design. But when I visited a highly favored site, I found one of the most idyllic landscapes in the Eastern United States—the kind of place I would fight to preserve. What troubles me still is that no one ever paid to add aesthetics into that early GIS, even though it would have been technically feasible.

During the years, I have used GIS to help make decisions with great impacts on people's lives, and the memory of that first experience always reminded me that GIS and its practitioners have limits. I doubt GIS will turn us all into "drive-by" developers, but we should question what GIS is doing to us and to society at large.

John Pickles (University of Kentucky), John O'Loughlin (University of Colorado), and several other geographers have expressed concerns about the ethical use of GIS and cultural representation in GIS. Tom Poiker (Simon Fraser University) has debated the impacts of geographic information and society. If you are inclined to view such deliberations as esoteric or unimportant, remember the reality of children killing children, and consider both sides of the ethical and cultural issues in GIS.

(This column originally appeared in the September 1993 issue of GIS World.*)*

GIS and Ethics: Out of Sight, Out of Mind?

By Nigel Waters

Delivering a keynote speech at a GIS conference is one of the most enjoyable tasks that I'm occasionally asked to perform. Thus I was delighted when John Thomas, a transportation planner with the city of Rochester, N.Y., and Jeff Volpe, a senior GIS analyst with Begrmann Associates, Rochester, asked me to give the keynote speech at the Tenth Annual GIS/SIG Conference. The GIS/SIG organization bills itself as a "forum for geographic information sharing that serves the spatial data and digital mapping community in the Genesee and Finger Lakes region of New York state." The only problem was that they wanted me to discuss the topic of "GIS and Ethics."

Online Resources

Information on GIS and ethics is conspicuous not by its absence, but rather by its relative scarcity. However, there are some useful online resources including those located at URISA's Web site (*www.urisa.org/ethics/code_of_ethics.htm*) and at the GIS Certification Institute (*www.gisci.org/code_of_ethics.htm*).

It can be argued that all such postgraduate courses should include such a module, and our own "Advanced GIS Course" at the University of Calgary now includes some lectures on ethical issues in GIS. But it also can be argued that such courses should become part of the certification process for GIS professionals.

The College of Geographic Sciences in Lawrencetown, Nova Scotia, Canada, doesn't appear to offer a course on ethical issues in GIS, but it's a topic of concern for at least one of their postgraduate students, Fiona Ryle, who posted an online paper about GIS ethical issues. The paper reviews three publications on the topic, and it's primarily concerned with matters of privacy and the illegal linking of spatial databases to produce new information. In most cases, jurisdictions in North America will only allow information to be used for the purpose for which it was collected.

In addition, a recent article by Karen Kemp and Richard Wright on "Educational Priorities for the University Consortium on Geographic Information Science" suggests that issues of access and equity should be treated as a priority in the GIS curriculum. They argue that information technology (IT) presents new opportunities for teaching those who have difficulties learning via traditional teaching methods. A previous "Edge Nodes" column suggested that IT also presents opportunities for sensory substitution, allowing those who have one sense impaired or lost to learn through the use of other senses (Waters, 1993).

Kemp and Wright suggested that a GIS Ethics Curriculum should be developed, including topics such as the true costs and benefits of GIS technology, legal issues, the varying degree of access to GIS technology and data, privacy, and how GIS is used to support decision-making processes. Case studies in a GIS Ethics course may include environmental justice cases, the identification and protection of land and resource rights, and cases involving public participation in GIS (PPGIS).

As part of its Varenius Project, the National Center for Geographic Information and Analysis, Department of Geography, University of California at Santa Barbara, produced a position paper on "Empowerment, Marginalization and Public Participation GIS" (*www.ncgia.ucsb.edu/Publications/Varenius_Reports/PPGIS98.pdf*). The paper, compiled by Will Craig, Trevor Harris and Daniel Weiner, set the research agenda for PPGIS and included abstracts of papers submitted on the topic. Abstracts featured such phrases as "methodology matters," "spatial empowerment," "Web-based GIS for public participation," "marginal societies and GIS," "marginalization," "legal access," "a community data system," "asserting new rights to know," "advocacy" and "disability." In addition, the abstracts frequently address ethical issues either directly or indirectly. (Full texts of the papers submitted for this project may be found on the Web at *www.ncgia.ucsb.edu/varenius/ppgis/papers/index.html.*)

Published Literature

Few of the standard introductory GIS textbooks—or even the more advanced works—address the issue of ethics at all. However, some textbooks discuss ethics under the categories of legal issues, privacy, equity and access.

Important exceptions to this relative dearth of published material on GIS and ethics include a seminal article by Will Craig (1993) and papers by Jeremy Crampton and Michael Curry (1995) as well as a more recent contribution by Nancy Obermeyer (1998). In Obermeyer's comprehensive article, she discusses four sets of responsibilities: 1) to the ideals of the profession, 2) between the profession and the client, 3) between the profession and society, and 4) between the profession and the student.

Most professions—especially those with certification processes such as engineering, law and accounting—frequently require some formal training in ethics. Computer professionals receive such training, and the Association for Computing Machinery has a code of ethics and professional conduct (*www.acm.org/constitution/bylaw15.html*). GIS ethical issues, with respect to certification as a GIS professional, are considered in an online paper by Peter Dale.

Obermeyer's second responsibility concerns the relationship between the profession and the client. In a ground-breaking article on GIS ethics, David Linden's concern that the proliferation of GIS companies and vendors would flood the market with low-quality products and over-hyped services proved unfounded. Instead, recent reductions in GIS software suppliers have led to a greater concern over market monopolization and the potential for predatory behavior by a few key players.

As far as the relationship between the GIS profession and society, there's little doubt that many communities are disenfranchised by the rush to advanced, computerized technology in delivering government services. (For an overview of government services via the Internet, see *www.netcaucus.org/books/egov2001/pdf/Ethicsin.pdf*.)

The last of Obermeyer's relationships is between the profession and the student. She argues that ethical concerns are receiving greater treatment in standard textbooks, but the discussion often comes at the end of a book and, thus, at the end of a semester. Therefore, ethics may not get covered in an increasingly overloaded curriculum, and professors may have little time for such discussions as the semester draws to a close. Obermeyer, citing the work of Waldo Tobler, notes that GIS theory and its associated concepts may have a "half-life" of 20 years, but technology is likely to have a half-life of less than five years (and perhaps nearer two). Ethical codes and guidelines, it might be hoped, would last much longer than technology, and the value of instilling these ideas would therefore be more beneficial.

The Digital Divide

A major concern in recent years has been the "Digital Divide," which was the topic of a paper session organized by Jeremy Crampton at the 2001 meeting of the Association of American Geographers (*http://monarch.gsu.edu/jcrampton/digital-divide/aag01.html#dha*). A primary research concern is to determine how income, education and race affect access to the electronic universe. In addition, there are many other variables that likely influence access and equity, including geography, age, physical disabilities, gender and psychological characteristics. Because of the "Digital Divide," it's possible that the Internet, rather than being the great leveler for information access, may have ushered in an era of unparalleled inequity.

Although the GIS/SIG Conference in Rochester didn't manage to address all the aforementioned issues, those in attendance should be grateful to conference organizers for making ethical issues "front and center," because they frequently have been ignored by the GIS industry.

(This column originally appeared in the August 2001 issue of GeoWorld.)

Think Twice About Kid-Tracking

By Jerry Dobson

In 2000, I warned about a future society monitored by Global Positioning System (GPS) technology and programmed through GIS (Dobson, 2000). At the time, the technology aspect was a prediction. Now human-tracking devices are openly sold, and still there has been no public debate about the ethics or efficacy of their use.

In 2003, Pete Fisher and I explored the potential for misuse and concluded that widespread "geoslavery" is inevitable (Dobson and Fisher, 2003). In this column, however, I set those horrors aside and address only the most benign application that has been proposed: kid-tracking by parents with good intentions.

In today's security-conscious society, many parents have accepted the "truism" that GPS-tracking devices can protect children. Yet remarkably little critical thought has gone into the question. Some flaws are obvious; others require a little thought.

To be clear, I'm talking about bracelets and implants now used to brand children with unique ID codes and report their location every couple of minutes (see Plate 8). I'm talking about the present generation of kid-tracking products that don't include transponders capable of remotely administering punishment. I'm talking about parental use that wouldn't qualify as geoslavery under the definition offered by Fisher and me.

Wireless Wonders

Kid-trackers use wireless networks, and wireless telecommunications are notoriously vulnerable. Hackers have learned how to convert Pringles cans into antennas for wireless signals (see *http://news.bbc.co.uk/1/hi/sci/tech/1860241.stm*). Last year at the University of Kansas, two graduate students, Matt Dunbar and Brett Becker, mapped all the wireless networks in Lawrence, Kan. (see *www.ittc.ku.edu/wlan*). They sampled the airways throughout town, easily capturing signals that gave all sorts of information about the senders. Few nodes bothered with encryption, and the encryption in place was based on standard commercial products so simple as to be practically useless.

This means any computer-savvy teenage prankster or adult predator may have access to the radio signal emanating from a child. If the service provider uses encryption, it only increases the level of expertise required of hackers. At an even higher level of expertise, the hacker may be able to break in on the communication and substitute bogus coordinates for the child's location.

Predators Apply Within

Kid-tracker signals don't go directly to parents. They go to commercial service providers, thus placing another human layer between child and parent. Most of the workers will be conscientious folks with good intentions, but how will employers discriminate against predators? And how will the entrepreneurs themselves be licensed, investigated and kept in check?

Kid-tracking companies will have power over children similar to that of daycare centers and schools, but they will not fall under the same boards and licensing procedures. In the United States, schools and churches are legally required to conduct background checks for prior sex offences. Computer companies are not. Predators already lurking on the Internet now perhaps can look forward to gainful employment.

Whenever a child goes missing, and especially when one is found alive, it's tempting to say, "If only she'd been wearing a locator device." The case of Elizabeth Smart sparked such sentiments, but what would those same people be saying if her abductor had owned the device? There isn't a single law or practice in place to keep an abductor from purchasing his own kid-tracker and using it to prevent escape.

Yes, clearly the technology can help if the abductor somehow misses it and allows it to remain on the child. I have no quarrel with that claim, but I'm troubled that no one mentions the adverse possibility.

Probing Questions

After my reservations about kid-tracking became known, several people expressed their opinions to me, and I've been struck by the lack of critical thought. The following are some examples:

Pro: Bracelets can be removed. Implants are the answer.

Con: A kidnapper will have no compunction about cutting a child. When he finds an implant, he will remove it, perhaps in the least humane way. If he suspects it's there, he may probe every scar and pimple until he finds it.

Pro: I wouldn't do it to my child, but I don't mind if other parents do it to theirs.

Con: The kidnapper can stop cutting the other child when he finds the implant. When will he stop cutting your child?

Pro: The only thing that matters is protecting my child.

Con: The vast majority of missing children are taken by family members, predominantly non-custodial parents. Indeed, the number of children taken by strangers is no more than 300 per year in the United States' population of 290 million. Thus, the person most likely to abduct your child also is most likely to share ownership of the device with you. If you're the custodial parent, make sure you get full custody of the device in the divorce decree and make sure you know how to run it at least as well as your departing spouse.

Victims of Love

I know a young man who is serving a life sentence for murder. We knew his family as neighbors even before he was born, and my wife, a teacher, knew them at school. He was the center of their universe, and they wouldn't have hurt him for the world.

His mother loved him so much that she often peeped through the classroom window to watch him in school. She watched him on the playground at recess and sat in the room during piano lessons. No school ever was good enough for him, so they moved him every few years, seeking one worthy of his genius.

Tennessee schools weren't good enough, so they moved to California, and, finally, a topnotch private college wasn't good enough, so they ordered him to his father's alma mater. For 18 years, they overprotected him and dominated every aspect of his life. Finally, he snapped and bludgeoned them to death. When neighbors back home in Tennessee heard the news, they understood. A child can only take so much control.

When I read about kid-tracking, the image I see is that overprotective mother peeping though her son's classroom window. That leads me to believe kid-trackers, on balance, will do more harm than good to the children they're supposed to protect.

I end here with three firm predictions: 1. More children will commit suicide because of these devices than ever will be saved from abduction by strangers. 2. More children will erupt in violent behavior against parents, siblings, friends and strangers. 3. More children, on reaching adulthood, will sue their parents and service providers for damages—mental abuse, emotional stress and psychological trauma—due to kid-tracking.

This is the Brave New World we have entered without public debate.

(This column originally appeared in the May 2003 issue of GeoWorld.*)*

GIS Is a Surveillant Science

By Nigel Waters

GIS has been shown to be a useful organizing technology because so much of the information recorded in memory aides has a spatial component. This column looks at criticisms of this type of technology and then briefly mentions some possible applications.

Future of Memory Aides

Much of EuroParc's research was anticipated by Norman (1992) who describes a memory aide that would be with us from the age of two. Norman referred to it as a "Teddy," because in some ways it would be a toy and companion for a youngster. The Teddy would know everything about us, store our experiences and provide access to the world's data banks.

Norman sees advantages and disadvantages to the Teddy. The advantages revolve around the fact that the Teddy is a useful help-mate. It will organize our lives far more effectively than the Personal Information Managers to which today's technophiles seem addicted. The disadvantages occur when the Teddy takes over. Norman himself discussed physical implants that would provide the necessary memory and interface. But he criticized Carnegie Mellon roboticist Hans Moravec, who speculated on how human memory might be transferred to silicon (Moravec, 1988; see Regis, 1990, for an unsympathetic review of Moravec's speculations). Perhaps the most chilling part of Norman's discussion is his ambivalent conclusion: "I have mixed feelings about this Teddy. I can imagine the good things. I can fear the bad. But in actuality, I have little choice in the matter. It is coming into being."

Criticisms of Intimate Technology

Mander (1991) provides a bitter denunciation of technology in general and computers in particular. He is especially lucid on the evils of keeping track of employees by computers. He cites the case of an information officer at Pacific Bell who has a computer monitor his performance from the moment he clocks on until the moment he leaves: "The computer knows everything. ... It's me and the computer all day. I'm telling you at the end of the day I am wiped out. Working with computers is the coal mining of the nineties." (Mander, 1991, p. 57).

More neutral is David Lyon's sweeping review of the rise of a surveillant society (Lyon, 1994). Lyon provided an introduction to the entire literature dealing with the problem of electronic surveillance. Particularly interesting is his use of Foucault's (1979) and Giddens' (1985) work and his ability to provide a conceptual framework within which to understand the pervasive influence of surveillance technology.

Foucault begins his study of prison reform with a gruesome account of torture from the 18th century. That is contrasted with Jeremy Bentham's Panopticon, a prison he designed that allowed the warden to monitor and control all prisoners from a central viewing tower. Each individually isolated prisoner "is seen, but he does not see; he is the object of information, never a subject in communication." (Foucault, 1979, p. 200). Foucault stated that the principles of the Panopticon can be applied widely: "If they are schoolchildren, there is no copying, no noise, no chatter, no waste of time; if they are workers, there are no disorders, no theft, no coalitions, none of those distractions that slow down the rate of work, make it less perfect or cause accidents." The Panopticon seeks to induce "a state of conscious and permanent visibility that assures the automatic functioning of power."

Like Lyon, Mark Poster (1990) extended Foucault's work into the computer age. He saw the development of modern database technology, in which consumers are the records, as a Superpanopticon and "a means of controlling the masses in the postmodern, postindustrial mode of information." Individuals today appear to be enamored of the new consumer-oriented surveillance technologies.

GIS: The Surveillant Science

It has been suggested that we can consider GIS as a science in its own right. If this is true and spatial databases proliferate, then perhaps GIS will become the surveillant science *par excellence*. GIS may be the first universal Panopticon.

Is there any good surveillance? Can EuroParc's research be used now in GIS with little to worry about? At the University of Calgary, there are several graduate theses under way that use electronic monitoring of large carnivores (specifically, wolves, black bears, grizzlies and cougars) for habitat analysis. EuroParc's software and GIS would make a perfect tool for that type of research.

(This column originally appeared in the October 1994 issue of GIS World.*)*

Are Robotics Laws Applicable to Location-Based Services?

By Daniel Sui

Although legal and social issues regarding the use of geographic information have concerned the GIS community for some time, such issues are far more urgent in the context of location-based services (LBSs) due to their potential ubiquity and close coupling with LBS users' personal lives.

LBS is the emerging technology that will test what Jerry Dobson (2000) in a previous "The 'G' in GIS" column called the ethical limit of GIS. Although calls have been made recently for a more comprehensive legal framework, no official legislation exists regarding the use of LBSs. Inspired by Isaac Asimov's "Three Laws of Robotics," Jean Michel Durocher (2002), CEO of Webraska Inc., formulated a set of "Laws of LBS."

Durocher contended that the LBS industry should adopt his three laws. This column provides a preliminary assessment on whether Asimov's three laws of robotics can be extended as a foundation for regulating LBSs.

Exploring the Laws

In anticipating the positive and negative consequences of Robotics, Asimov presented his laws of robotics (below) in a short story called "Runaround" released in 1942.

Since their publication, Asimov's laws of robotics have been continuously debated and fine-tuned. Although still controversial, they have been widely accepted as a reasonable guide to discuss the moral issue of robotics. Even Asimov admitted that "there was just enough ambiguity in the three laws to provide the conflicts and uncertainties required for new stories."

Although more fine-tuning is needed, I think Durocher's laws of LBS (below) offer a reasonable guideline for LBS use in the near future—provided that an enforcement mechanism can be established.

The first law addresses the requirements of LBSs in emergency situations and is consistent with enhanced 911 services as mandated by Federal Communications Commission regulation. It's reasonable to expect improved emergency services if all 911 calls are locationally explicit. The first law of LBS also requires that no services should put mobile users in danger.

The second law ensures that LBSs are voluntary, and privacy should always be preserved for individuals subscribing to these services. Under normal circumstances, a subscriber should be able to prevent others from locating him or her. But totally complying with this could put users at risk, because, in emergency situations, it might be impossible for users to provide all the necessary life-saving authorization, and, in some cases, the police will want to (and already do) track cellular phone use to locate dangerous criminals.

The third law stresses that location alone, isn't sufficient. Service providers must find a way to develop the proper business models and charge for the proper services. Applications must bring real value that consumers and business users will be willing to pay for. However, the third law also ensures that no business should be able to benefit from the availability of location data to the detriment of an LBS user's right to complete privacy and safety.

Asimov's Laws of Robotics		Durocher's Laws of LBS
First Law of Robotics: A robot may not injure a human being or, through inaction, allow a human being to come to harm.		First Law of LBS: Location, through its availability or non-availability, must not allow a human being to come to harm.
Second Law of Robotics: A robot must obey orders given to it by human beings, except where such orders would conflict with the First Law.		Second Law of LBS: The availability of one's Location must be in one's complete control, except where such control would conflict with the First Law.
Third Law of Robotics: A robot must protect its own existence as long as such protection does not conflict with the First or Second Law.		Third Law of LBS: The providers of location-based services must be allowed to create a profitable business from these services as long as such business does not conflict with the First or Second Law.

Seven Scenarios

As Peter Fisher and Dobson (2003) illustrated in their seven different scenarios regarding the use of LBSs (see accompanying table), the legal and ethical implications of LBSs are complex and contingent on many uncertain parameters in the real world.

Surveillance Scenarios and Laws of LBS			
Scenarios	Examples	Ethical Issues	Laws of LBS
They didn't know where you were	Normal life for ordinary citizens	Personal privacy	Obeyed (violated if the person is suicidal?)
They know and you want them to	On call, emergency, alibi	Consent	Obeyed
They don't know, but you thought they did	Faulty signal, signal blocked, physical density	Ignorance	Violated
They know, and we wanted them to	Criminal activity, unethical conduct	Public interest	Ambiguous
They know, and you didn't want them to	Nuisance calls, stalking, monitoring	Geo-identity theft	Violated
They know and we thought we wanted them to	Abusers, hostage takers	Coercion	Violated
They know and can physically keep you there	Forced labor or domestic abuse	Geo-slavery	Violated

Nonetheless, the three LBS laws provide reasonable guidelines for most civilian applications during peacetime, but the three laws run into moral and ethical difficulties in the context of war, especially in cases like the ongoing war on terrorism.

Must all three laws be necessarily violated to win such a war? Or will there be permanent peace on Earth after all three LBS laws are strictly observed by each and every individual citizen of the global village?

(This column originally appeared in the September 2004 issue of GeoWorld.*)*

GIS and Martus: Technology with a Conscience

By Nigel Waters

I have argued earlier that GIS technology can be used for altruistic purposes as well as to serve an unjust status quo or facilitate the growth of a more repressive social, economic and political environment. Subscribing to a view that technologies such as computers or GIS are "good or evil" in and of themselves seems naïve, but in the early days of quantitative geography, many commentators did just that, and, more recently, GIS has been targeted with such criticism (Pickles, 1994).

Therefore, I'm always delighted to discover humane and compassionate uses of computer technology that could be combined with GIS. Martus would appear to be one such application.

Martus Security

Palo Alto, Calif.-based Benetech manufactures Martus software (*www.martus.org*), which acts as a human-rights bulletin system. The word "martus" means "witness" in Greek. The free software enables those living in parts of the world where human-rights abuses are common to record the abuses and attach supporting documentary evidence.

Martus users can create secure, private bulletins that contain the following information: language used, author name, organization, title, event location, keywords, date the event occurred, date the event was recorded, summary, additional details, attachments such as photographs, and other documentary files.

After creation, the bulletin can be uploaded to the Martus servers located at various locations, and the information is encrypted and backed up. Martus is an open-source application, so programmers and users who may be wary of the software can view the code.

GIS and Martus

GIS files may be added to Martus bulletins as attachments. The simplest GIS use would be to show where events and human-rights abuses occurred. If someone had been shot, for example, the location of killer and victim could be indicated along with the line of fire.

In Canada, and presumably most Western democracies, there's a right to demonstrate peacefully, and this includes the right to be seen. Therefore, visiting foreign dignitaries aren't protected from seeing protestors who disagree with their human-rights records. GIS could be used to document the dignitary's viewshed, and the software could determine if they could see the protestors.

This isn't a hypothetical example. In 1997, President Suharto of Indonesia was visiting Vancouver, British Columbia, Canada, for the Asia Pacific Economic Cooperation (APEC) Conference when an incipient and peaceful protest was violently dispersed by police action. The final report into the alleged mishandling of the protestors by the Royal Canadian Mounted Police (RCMP) wasn't released until 2002.

Commenting on the report, Eric Tolpannen, a crown attorney for the Alberta Justice Department, noted that sites for the demonstrators were too small, and buffers between protestors and dignitaries were so large that the protest would have been ineffectual. Using GIS, both requirements could have been determined ahead of time.

Tolpannen suggested that the lessons learned at the APEC Conference could have been useful for the G-8 Meeting that was held later in 2002 in Kananaskis Country, Alberta, Canada. As events transpired, the G-8 Meeting was one of the quietest ever. The leaders landed in a remote spot at the Calgary airport and were whisked by military helicopters to an isolated forest setting for the meeting. No protestors were allowed near the leaders. You didn't need a GIS to show that the rights of those wishing to protest weren't considered.

Interestingly, Calgary municipal officials advised authorities in Savannah, Ga., on how to host the G-8 summit on June 8-10, 2004, in a relatively remote and defensible location. Protest Web sites already are online (e.g., *www.repealnow.com/g8.htm*), but the protestors in Georgia weren't able to get any closer to the delegates than those in Alberta in 2002.

Martus and Juliet O'Neill

For those of us living in a Western democracy, freedom of the press is essential. It was particularly disturbing, therefore, when on Jan. 22, 2004, the home of Juliet O'Neill, a reporter for the Ottawa Citizen newspaper, was raided by the RCMP.

Police officials were looking for the identity of a source that had allegedly leaked details of a security dossier on Maher Arar, the Syrian-born Canadian citizen who was deported by American authorities (allegedly with Canadian intelligence service support) to Syria, where he was allegedly tortured. O'Neill said her address books, contact book and Rolodex were taken, and her computer hard drive was copied (see *www.nytimes.com/2004/01/22/international/americas/22CANA.html*).

Now, can you see why we need Martus? And GIS support files wouldn't hurt.

(This column originally appeared in the March 2004 issue of GeoWorld.*)*

GIS and the Bitter Fruit: Privacy Issues in the Age of the Internet

By Nigel Waters

A relatively new development in the GIS world is the increased use of the Internet as a processing platform. In the future, organizations may not actually own an expensive GIS package. Instead, they may simply submit data to a GIS vendor and have the results downloaded after paying a transaction fee.

A few weeks ago, for example, I wanted to purchase a site license for Chen and Getis' Point Pattern Analysis program. When I e-mailed one of the authors, Art Getis from the San Diego State University Geography Department, he informed me that I didn't have to buy anything. This useful spatial analysis program now was available on the Internet. All my students had to do was visit the Web site where the program resided, upload their data and "presto," the analysis would be carried out for them and downloaded back to their computer. It worked like a treat.

In this instance, we were uploading a minimal set of information—a previously published dataset. Suppose, however, we had been working with a GIS vendor. And what if we were uploading a complete set of GIS coverages? We then may have been entrusting sensitive information to a third party. Moreover, if we were to do this on a regular basis, we would provide vendors and Web site owners with a profile of our activities, and that may be a less attractive side effect of this convenience.

An Increasing Concern

Privacy always has been a concern on the Internet. Matt Richtel's column in the *New York Times* on March 12, 2000, marveled at the fact that New Yorkers are quite willing to divulge to e-commerce businesses all kinds of personal information that we would be horrified to tell a stranger. Yet after the information has been passed to an Internet company, such data can be used in many ways that are much more intrusive than had we divulged our personal characteristics to a stranger. The mechanics of online data gathering from Web site visitors are discussed and described in a series of articles by *Smart Computing* that may be found at *www.smartcomputing.com.*

The Future of Privacy

Michael Sweet (2000) cites a Georgetown Internet Privacy Policy Survey, which found that almost 91% of the commercial Web sites surveyed collect e-mail addresses, 81% collect names and almost 63% collect postal addresses. In the "good guys' corner" are organizations such as the Direct Marketing Association and TRUSTe (*www.truste.org*). These organizations help online businesses institute privacy policies, but, as with the Better Business Bureau, there's no legislation forcing e-commerce businesses to comply.

Sweet notes that although individuals may do much to protect themselves from unwarranted intrusion by Web site owners, this isn't necessarily the case with their own employers. In 1999, 17% of Fortune 1,000 companies were monitoring their employees. By 2001, this number is expected to have risen to 80%. Sweet also refers to another study showing that the number of major U.S. companies checking employees' e-mail rose from 15% to 27% between 1997 and 1999 and, during the same period, those checking computer files rose from 14% to 21%.

Online Profiling

Advertising only works well if you can be sure of your target audience. Because online advertisers spent about $4 billion in 1999 (Farwell, 2000), they want to make sure they're getting value for their money. Advertisers can do this by having the Web sites on which they advertise track their customers via "cookies," which are text files added to a user's computer when he or she visits a Web site.

A cookie allows Web sites to identify customers when they next visit the site. The cookie can store personal information about the customer as well as which links the visitor clicked on, allowing the site to greet customers on their next visit and show them only the advertising they may be most interested in—as indicated from their past history of clicking.

If you've ever bought a book from Amazon.com and revisited, you'll know all about this. The Amazon Web site will tell you which other books were purchased by individuals who purchased the book you just examined. The real danger is that this type of information can be linked to other Web site cookies and databases. Case-based reasoning and similar software then could be used to build detailed profiles on individual users.

Fighting Back

Perhaps you're not worried about who knows what you're reading. Perhaps you don't mind uploading credit card numbers and datasets to GIS vendors. But if you're concerned, then standard browsers such as Internet Explorer and Netscape Navigator allow you to turn off your computer's ability to accept cookies and delete existing cookie files. Jennifer Farwell (2000) notes that the Junkbusters' Web site (*www.junkbusters.com/ht/en/cookies.html*) provides cookie management software and a link to a site that will determine what information is being divulged by your browser.

This column concentrates on how to protect oneself when surfing the Internet, registering at an Internet site or uploading information. Many individuals, however, have their own Web sites, and, if you gather information on other people, you should maintain your own privacy protocol. Most universities today have ethics committees that review grant applications that relate to human subjects. What's not so common is for universities to have protocols for privacy protection of individuals who visit Web sites belonging to university personnel.

Eileen De Mott Curtright (2000) provides guidelines on what to include in such a privacy statement. She also describes a case in which the U.S. Federal Trade Commission (FTC) charged GeoCities with "unfair and misleading business practices after finding that the site misrepresented the way personally identifiable information it collected through its membership application was used."

GeoCities, which settled with FTC in 1998 without admitting any wrongdoing, was subsequently acquired by Yahoo!. If you want to see an FTC-endorsed privacy statement as a model for your own Web site or what you have a right to expect from other Web sites, go to *http://privacy.yahoo.com*.

The Transparent Society

In his recent book, David Brin (1998) argued for a "transparent society" in which everyone knows everything about everyone else. He cites the fable of the blind society in which only a few who were willing to eat an exceedingly bitter fruit could see. After a while, the people with sight began to abuse their power and take advantage of the blind. The solution was for all to eat the bitter fruit so everyone could see everything and prevent this abuse.

For Brin, the bitter fruit we must swallow is the realization that we're all responsible for keeping an eye on the world. He argues that we will prevent abuse of surveillance technologies such as cookies and online profiling not by government regulation, but rather by a mutual, shared responsibility for the world in which we live as well as a disdain for those who abuse and misuse this privilege. Many others may feel that ubiquitous surveillance, whether in real or virtual worlds, is a bitter fruit indeed.

(This column originally appeared in the May 2000 issue of GeoWorld.*)*

Are the Rich Getting Richer and the Poor Getting Poorer with GIS?

By Nigel Waters

With GIS, one may legitimately ask the question, "Are the rich getting richer, and the poor getting poorer?" Is GIS a technology with a liberalizing, democratic trend to it or a technology that will concentrate power and information in the hands of the fortunate few who have the money to buy costly computer hardware, software and information systems?

Is Information Power?

Some of these concerns already have been aired on the GIS-L electronic forum in a rather short-lived and inconclusive debate. Related issues have been addressed in the popular news media. J. Coates (1992), referring to electronic retrieval systems, notes, "What was heralded as an information renaissance may be a new dark age, complete with a class of data lords and one of dataless serfs." Coates argues that the problems began to emerge around 1985 when North America went online, and newspapers began selling their copy to electronic databases. Today, all of this information is at your fingertips, but only as long as you have the necessary computer equipment, the training, the ability to use it online and the money to pay for the connection charges.

Coates estimated that it probably would have cost about $1,780 to download the texts of all 636 articles that he wrote for the *Chicago Tribune* since 1984 when that newspaper first began selling its material electronically. Of course, the person who obtains this information in electronic form is presented with enormous power for carrying out searches and content analysis of the stories, capabilities that would not easily be available to anyone who had, instead, laboriously extracted this information from microfiche or clippings over the years. In the past, most politicians subscribed to clipping services, but they would be much more effective if they could afford to obtain, monitor, analyze and manipulate the clippings electronically. The trite adage that information is power is valid and, as Coates has remarked, management information systems executives who control data access are jokingly referred to as Givers Of Data—the acronym should be obvious.

Many of the technological developments of the industrial revolution appear to have had the desired liberalizing tendency (Snow, 1959). Thus, the printing press introduced the possibility of universal literacy, and developments in transportation led to mobility for the masses. But the models we have for generalizing the development process of transportation systems seem to suggest otherwise. The best-known model of the development of such systems was provided by E. Taaffe, R. Morrill and P. Gould (1963), demonstrating how a transportation system might evolve from a non-industrialized state into a modern network. But with each technological development, such as railways, super highways, airline systems and high-speed rail services, the number of nodes and links in the network favored with access to the new systems became fewer and fewer. The cost was too great to incorporate each of the nodes that were part of the original trail system. In Canada, only one set of links, those in the Quebec-Windsor axis, may ever have the population density to justify introducing specialized, high-speed rail service.

The transportation model is an old one—an extension of the venerable central place theory, which states that high-level services are provided only at a few privileged locations. Similar models have been used by economists such as Nobel Laureate Gunnar Myrdal (1957) to explain regional inequalities. Transportation specialists and economists argue for a world in which cumulative causation and positive feedbacks—the centripetal effects—override the negative feedback centrifugal checks that tend to even out growth and spread economic advancement among the have-not regions.

But isn't information different? Aren't fiber-optic networks so inexpensive that information will be provided to all? Pending some positive election results, won't the promised land of visionaries such as Al Gore (1991; see also the summer 1991 issue of the *Whole Earth Review* for articles by Howard Rheingold and others on "Electronic Democracy") mean that the transportation and economic growth models are irrelevant for GIS and information systems generally? Perhaps. But in GIS, some trends seem to be moving in the other direction.

Although hardware is becoming less expensive and we are getting more million instructions per second (MIPS) for our dollar, it never seems to be enough. The swap space for running some commercial GIS packages is now recommended to be more than 100 megabytes; for some reason 32 megabytes or more or RAM is useful; workstations, preferably big ones, are preferred over PCs for doing "real work;" and the hardware and software you bought last year is now embarrassingly out-of-date and so are the data. Big departments in big institutions, big governments and big universities with big budgets are favored over lesser organizations.

While ideology promotes the spread of technology in North America, economics may still prove to be a hurdle. As N. Weinberg (1990) argues, comparisons with the telephone system also are illuminating. Simply put, "The [economic] advantage that may be gained from access to persons or information leads us to sequester and hoard these resources."

(This column originally appeared in the November 1992 issue of GIS World.*)*

GIS Technology Reveals "Revenge Effect"

By Nigel Waters

Edward Tenner's (1991) *New York Times* article on the "revenge" effects attributable to our technological ingenuity concerned such things as why plane seats seem to get smaller as planes get larger and why voice mail seems to double the time it takes to complete a phone call. He felt that these revenge effects, in which our cleverness is twisted against us, could be slotted into five categories: repeating, recomplicating, recongesting, regenerating and rearranging. In many cases, revenge effects appear to have manifested themselves in GIS use.

According to Tenner (1991), repeating occurs when "a task is made easier or faster but becomes required more often." He cites the use of spreadsheets, but perhaps GIS professionals can note how GIS technology has increased the demand for maps and other forms of spatial information. When a map, like a spreadsheet, was difficult and time consuming to produce, it was requested rarely and only when it was really needed. Now demand is more frequent, and supervisors may insist that a map be revised again and again with trivial changes—and why not, since it is so much easier to do.

Recomplicating suggests that technological advances such as voice mail often make simple things more complicated and difficult to do, and may even require that personnel take retraining courses. The skills required to access spatial information are more sophisticated than in the old days when data could be obtained from a filing system. For example, many fire departments now use computer-aided dispatching systems (a form of GIS), but they also maintain their running orders on an old-fashioned filing system, if only to prevent a total loss of system functionality with a computer crash. The replacement of paper road maps with onboard navigation systems for cars is another example of recomplicating a previously simple and straightforward task (see Plate 9).

Recongesting occurs when "an updated function becomes slower and less comfortable than the original," said Tenner. Travel may become easier due to the construction of a new road, but then traffic gradually increases until the new road is as congested as the old road. Tenner quoted philosopher Ivan Illich, who estimates that Americans spend an average of 1,600 hours each year working to support their automobile to cover an annual commuting distance of 6,000 miles. In other words, for all their efforts they only have achieved a good walking speed. Perhaps a similar equation might be developed for expenditures on computing equipment. Sitting in front of a monitor all day can have unpleasant side effects.

With regenerating effect, a problem appears to have been solved, but the solution revives or amplifies the original difficulty. GISs often are used for pest control. A video from ESRI shows the use of GIS to identify mosquito sites, but, as Tenner noted, spraying may kill other species that feed on the original pest.

Finally, there is rearranging. Here the problem may be delayed, but when it eventually occurs, perhaps in another location, the result is worse than if the original situation had been left to run its course. Tenner cited John McPhee's 1989 book *The Control of Nature* in which it is argued that attempts to control minor floods in California's San Gabriel Mountains may lead to more massive flooding. Similarly, the use of GIS to prevent wildland fires (Bradshaw, et al., 1987) may lead to a dangerous build-up of litter, which might result in a more devastating burn later.

Of course, there are benefits to technological developments, and Tenner admits this. But what he does not resolve is why people bother with technological progress when these developments have so many negative side effects. Perhaps the answer lies in the work of the Danish economist Ester Boserup. Boserup is widely regarded as having "turned Malthus on his head" (Schultz, 1990) in suggesting that it is population pressure that has necessitated technological change. She argues that population pressure has demanded more intensive forms of agriculture and regards population growth as the independent variable (Boserup, 1965) and not as the Neo-Malthusians contend, a variable that depends on agricultural output. Boserup's work is largely confined to agricultural societies and those in the earliest industrialization stages, but it may have implications for technologically advanced societies and give new meaning to the old adage that necessity is the mother of invention.

(This column originally appeared in the November 1991 issue of GIS World.*)*

Company Profiles Make Business Sense to GIS Students

By Daniel Sui

The commercial sector has played an increasingly important role in developing GIS technology. The emergence of a mature GIS industry has received much attention by the business community, as evidenced by the coverage of GIS by magazines such as *Fortune*, *Forbes* and *Business Week*. GIS as a booming business, however, receives little attention in higher education. Most of the current GIS curricula are confined to the dichotomy between education (teaching concepts) vs. training (teaching of specific software packages) (Kemp et al., 1992).

I believe that educating GIS students from a business perspective is a new challenge for educators. To prosper in today's exceedingly competitive job market, it is important for GIS students to understand how the GIS corporate arena works.

A New Approach

In 1993, I introduced a new mini-project into my undergraduate GIS class at Texas A&M. The exercise gives students a business perspective without interfering with the education and training part of the class. This project is simple and easy to implement. I ask each student to develop a GIS company profile using the following steps:

- Find a company using resources like industry magazines such as *GeoWorld* magazine or portals such as GeoPlace.com.
- Get necessary literature and material, which may include documents about the company, product brochures, product demonstration software, etc.
- Write a 500-word report about the company and give a five-minute oral presentation to the class.

I require the written and oral presentation to cover a brief history (how the company got started), major products/services (what kinds of hardware, software, data and consulting services are offered), major market segments (who are the major customers), employment requirements (what kind of employee the company is looking for) and problems/difficulties in running a GIS business.

I intend to accomplish two goals with this mini-project. The first is to build a dialogue between the GIS business world and the college classroom. This exercise gives each student an opportunity to explore the GIS corporate world by learning employment requirements and the latest products. The second goal is to train the students to develop good oral and written communication skills. Based on the results of two semesters, I believe the two objectives have been achieved.

About 30 students enroll in my GIS class, which means each student is introduced to 30 companies during the oral presentations. The students feel that a GIS is more than a combination of hardware, software and data. They say the company profile exercise makes the GIS industry real to them. In fact, students who profiled GIS companies in the Houston and San Antonio areas were invited to tour the companies. Some were offered internships and jobs from the companies they profiled.

Students also say that this simple exercise makes a lot of business sense to them. They told me that by talking to the people in the GIS industry, they have a better idea about where GIS technology is going and the skills the industry desires. That helps them get their education plans more attuned with today's job market. During the profile process, students also get a chance to practice writing and speaking professionally.

I believe the success of an undergraduate, introductory GIS class depends on a clear understanding of concepts and sufficient training with specific software tools, as well as a deep appreciation of how the GIS corporate world works. Today's job market demands that students are intellectually and technically competent, and keen in business sense. A company profile is an easy, effective way to achieve this goal.

(This column originally appeared in the March 1995 issue of GIS World.*)*

Do We Need a GIS Certificate?

By Jeremy Crampton

There's now a nationwide GIS Certificate. Established in January 2004, it's run by the GIS Certification Institute (GISCI), a nonprofit spin-off of the Urban and Regional Information Systems Association (URISA). To be certified, an applicant must submit the $250 application fee, demonstrate sufficient experience in GIS and sign its Code of Ethics. If they qualify, they become a GIS Professional (GISP).

Developing the certificate has been a long process, involving a lot of hard-working people. But not everyone agrees that it's beneficial.

According to Francis Harvey, a professor at the University of Minnesota, "achievement of skill-level, point-based certification approaches such as that being promoted by URISA ... can become so nonspecific as to be meaningless for employers looking for concrete measures of job candidates' skills and abilities." Harvey (2003) concluded that such vagueness makes the certificate meaningless by itself.

Suggested Benefits

GISCI lists several benefits of the certificate, including that it "maintains and protects a core body of knowledge." Aspiring GISPs get points for coursework or degrees that cover one or more of 12 areas of the Geographic Information Science and Technology "core body of knowledge." These areas include "conceptualizations of space," "data modeling" and "geocomputation" (UCGIS, 2003). A candidate doesn't need all of these—they can qualify by having a bunch of classes in just one area.

As a cartographer who sees GIS as part of mapping and not the other way around, it's particularly chastening and dispiriting to see "cartography and visualization" being given such a small "walk-on" part. GISCI should read Keith Clarke's textbook, *Getting Started with GIS (Fourth Edition)*, which shows how GIS has its roots in cartography.

But what worries me most about this core body of knowledge is that it seems entirely oblivious of the critical GIS perspective that has emerged during the last decade and a half. The "core body of knowledge" that informs the GISCI Certificate is almost completely focused on technical and scientific aspects of GIS.

Indeed, the document begins with the bald assertion that GIS is a science: "We reject the notion that GIS is merely a collection of tools and techniques. Rather ...

[it is] in some of its aspects, a science." Its other aspects, the report continues, are more like engineering (UCGIS, 2003).

The final core body of knowledge includes "professional, social and legal aspects," but this turns out to mean legal and administrative issues. A critical GIS isn't just something you throw on at the end, but rather something that informs mapping and GIS from the beginning. Therefore, I'd suggest that critical GIS be a compulsory part of the points system of accreditation.

Know the Code

To get a GISCI Certificate, applicants must sign a Code of Ethics. In fact, the code came before the certificate and was being discussed as early as 1993 by Will Craig. At that time, Craig argued that a certificate based on training was worse than a code of ethics that has to do with how someone performs.

Craig also argued strongly for enforcement and sanctions for code violations. As far as I can see, however, there are none in the present certificate. Sanctions in other organizations include private and public censure as well as being disbarred from membership.

The Code of Ethics is a central element to GIS Certification.

The GISCI code suggests that those certified be objective and "not duly swayed by the demands of others," but what does this mean? Objectivity isn't what's needed during situations that involve ethics. Instead, people should assess vested interests and then judge whether their job is at risk. Can they get something done, or will powerful authority figures block them or colleagues see them as a troublemaker?

If there's to be a realistic code of ethics, then it should emphasize unequal power relations and that there are centralized hidebound bureaucracies that constantly claim "it's always done this way" or don't care about sanctions.

A better code would need to recognize the "real-world" context in which people work. But I don't know if any code can do this—they're internal and inward-looking by nature. And having one gives a false sense of problem solving.

As Craig admits, a code of ethics also allows a profession to be taken more seriously as a mature field. So having a code is also about gaining benefits for the profession—it's a social document.

Serious Concerns

The GISCI Certificate and the Code of Ethics seem to me to be seriously deficient. I'm not opposed to either in principle, but a certificate that baldly asserts the irreproachable "scientificity" of its subject, and a Code of Ethics that lacks sanctions and acknowledgement of the power structures of GIS knowledge are, to my mind, divorced from reality.

(This column originally appeared in the April 2005 issue of GeoWorld.*)*

Zen and the Art of GIS Maintenance

By Nigel Waters

The main thrust of Pirsig's 1975 cult classic was to know your machine. He had nothing but contempt for his friend who preferred to let a competent mechanic maintain his motorbike. Pirsig believed in gumption, an old Scottish word meaning common sense or a practical shrewdness. People who have gumption can fix their own bikes.

Pirsig is not alone in admiring this quality. Pagels (1989) pays tribute to the Japanese for their "bottom-up" approach to the understanding of technology. "The hands-on approach is the key to success in modern science; people who don't want to get their hands dirty have no business in science... I am disturbed by the tendency of some educators to promote a top-down approach."

And what does all this have to do with GIS? Well, isn't this the way many institutions are teaching GIS? Is it not, to a large extent, a button-pushing exercise? Do the students running a GIS package really understand the statistical routines they are running? Do they ever get to see the code?

Under the Hood

In many cases they are using proprietary packages whose owners do not wish to reveal the inner workings of their products and this is understandable. But the software developers could provide equations and pseudo-code for the various algorithms they are using. Procedures less important to the user, e.g., how the code was made more efficient and optimization techniques, are what developers might legitimately wish to keep secret.

The ability to interface your own code is one development in the spirit of Pirsig's do-it-yourself mentality. We want users aware of the precise nature of the transformations, interpolations, weighting procedures and a host of other data modifications carried out inside the black box. The user who knows exactly what happens to the data from input to output and can add modules to the package and modify it when it does not produce the desired results is a user with true gumption.

Overcoming Barriers

Pirsig (1975) talked about gumption traps. Some arise from external conditions and some are internal, arising from the individual. These do occur in the GIS world. Overlays represent a simple operation that must be carried out in the correct sequence. Things that work sometimes but not at others are the bane of consultants.

The right components also are a concern in a GIS set-up. These are problems software developers might do something about. Making it impossible to carry out illogical sequences of operations would help to solve the first problem.

Internal gumption traps must be addressed by the educator and include: value traps, such as rigid thinking, ego (common in computer "experts"), anxiety, boredom and impatience; truth traps, such as Pirsig's *my* state, which means that the answer to the problem is beyond the original hypothesis and cannot be answered by a simple yes or no; and muscle traps, such as not using excessive force (it applies to hung RAM as well as rusted screws).

One necessary quality is the willingness to experiment, to play with the machine or the computer or the GIS.

How much do we really need to know about the mechanics of GIS? The more the better. When Intel's 486 chip had some malfunctions, one article described them as obscure trigonometric functions. But in some graphics procedures, such as rotations, such functions might be fundamental. Knowing this level of detail might be important, although a casual decoupling (Pagels, 1989) in the chain of explanation might represent a convenient break in the level of knowledge.

Pagels (1989) argues for the importance of new instruments in the progress of science: "Theoretical ideas follow after the instruments and experiments; and these theoretical ideas, in turn, lead to the construction of new instruments." GIS can be seen as a new set of instruments leading to new theories and further technological refinements. But for this we need to know what is going on inside each black box.

(This column originally appeared in the February/March 1990 issue of GIS World.*)*

Can GIS Help Reverse the Decline in Boys' Grade-School Performance?

By Nigel Waters

Recent research suggests that girls outperform boys in grade school. It also has been suggested that GIS can help the boys' cause. Can this be true?

To prove that GIS *can* help boys' grade-school performance, it's necessary to show the following: 1) boys' grade-school performance has been declining relative to that of girls during the last decade or so, 2) this is unfairly jeopardizing their education, 3) boys have better spatial abilities than girls, 4) boys are attracted to the high-tech wizardry of computer games and virtual reality, and 5) such a predilection for computers will transfer to computer-based learning environments such as GIS.

A case can be made for the above scenario, but it's far from simple. If the chain of logic is valid, then it can be argued that although GIS unfairly favors boys, this is good, because it may reverse other educational disadvantages they're experiencing in our schools.

A Decline in Performance

The debate about whether boys were being shortchanged in schools came to a head in 2000 with the publication of Christina Hoff Sommers' book, *The War Against Boys: How Misguided Feminism is Harming Our Young Men*. The book was a reaction to a report by the American Association of University Women (AAUW) released in 1992 and published as a book in 1995, *How Schools Shortchange Girls.* In 1998, the AAUW published a second study, *Gender Gaps: Where Schools Still Fail Our Children,* which also was largely unsympathetic to the plight of boys.

Judith Kleinfeld, writing in *The Public Interest* (1999), provided a conservative ideological view of the debate. She noted that in the early part of the 20th century, the concern was all for the mistreatment of boys in school. But by the end of the century, this concern had changed and was overwhelmingly for girls. Kleinfeld quotes one teacher as stating, "I know that boys have problems. But we just don't give the boys much attention." Kleinfeld concluded that "teachers are honing in on the problems of girls, and they are overlooking the problems of boys."

The decline in boys' grade-school performance has been noted in several western countries for several years. For example, in 1996, *The Economist* ran a cover story on the topic. Based on research by the Surrey, U.K.-based Educational Psychology Service, 5- and 11-year old girls were shown to outperform their male counterparts in English *and* mathematics.

The U.K. Equal Opportunities Commission showed that 16-year-old girls had outperformed boys from 1985 to 1994 in English in public examinations. The report showed that boys outperformed girls in mathematics throughout the time period and in computer studies up to 1988, when the performance of girls became superior and remained that way until the end of the period.

The latter observation is bad news for our present argument, but the gender gap here was much smaller than in English, and, of course, we're also arguing that boys traditionally have been seen to have better spatial abilities than girls do. But is the statement regarding spatial abilities still true? Was it ever true?

Boys' Spatial Abilities

The literature on whether boys' spatial abilities are better than girls' is equivocal. Because the literature is extensive, inconclusive and sometimes downright contradictory, what's a researcher to do? One approach to this conundrum is to carry out a "meta-analysis."

Meta-analysis became popular when Gene Glass invented the term in 1976 (Glass, 1976). It's a methodology that allows the averaging of statistical measures such as correlations (although many other statistics may be used) across studies. It has proved especially popular in medicine, psychology and education. Occasionally, it also has been used in GIS studies or studies that might be facilitated through the use of GIS (for example, see the Global Change Ecology work at the Stockholm Environment Institute on the Web at *www.york.ac.uk/inst/sei/seiyinfo.html*.

It's the application of meta-analysis in the field of educational psychology with which I'm concerned here—specifically the magnitude of sex differences in spatial abilities. The likelihood is that there's a continuum in spatial (and other abilities), and there's some evidence that this may be due to fetal exposure to androgens such as testosterone.

So what does the meta-analysis say? According to Daniel and Susan Voyer and M.P. Bryden (1995), an almost exhaustive analysis of 286 studies from the psychological literature from 1974 to 1993 clearly established that sex differences in spatial abilities were statistically significant in favor of males. However, the differences varied by test used, declined over the period of the study and emerged between the ages of seven and 14 years old. So this is good news for the boys.

GIS and Virtual Reality

Two recent books are dedicated to the topic of teaching GIS in a school environment. The first is David Green's *GIS: A Sourcebook for Schools*, and the second is Richard Audet and Gail Ludwig's *GIS in Schools*, but neither mentions the possibility of using GIS to redress the recent decline in boys' school performance. This also is true of Sarah Bednarz's (1995) paper on the teaching of GIS using a constructivist.

So there seems little interest in helping to improve boys' performance by piquing their interest with GIS technology's technical wizardry and by taking advantage of their apparently superior spatial skills. However, if GIS embraces virtual reality, this may all change.

Virtual reality and GIS would seem to be a natural fit, and this would appear to be the view of those working at the Centre for Advanced Spatial Analysis (CASA) at University College London, where several projects are merging the technologies to create 3-D models of cities (*www.casa.ucl.ac.uk/newprojects.htm*). The models use Virtual Reality Modeling Language and, in the future, may also employ Geography Markup Language.

Some research suggests that the gender bias is even more markedly in favor of boys in the world of virtual reality (see *www.trnmag.com/Stories/070401/Gender_gap_shows_cyberspace_bias_070401.html*) and that this will need to be addressed to avoid discrimination against women in computer-simulated training environments. This implies that the virtual reality/GIS interface truly will be an environment in which boys may shine, but the evidence is equivocal. Albert Rizzo of the University of Southern California, for example, found no gender differences in spatial rotation tests in a virtual reality environment (see *www.sciencedaily.com/releases/1999/03/990318040810.htm*).

Therefore, as soon as enough virtual reality studies are available, I'll need another meta-analysis to determine if gender differences truly exist. Until this occurs, perhaps not even GIS will speak for the boys.

(This column originally appeared in the August 2002 issue of GeoWorld.*)*

Should GIS Stay Out of Schools?

By Nigel Waters

My most vivid memory of February's GIS '97 conference was the end of the GIS in Education session. A representative from ESRI Canada Ltd., North York, Ontario, Canada, explained that Canadian schools could acquire an ArcView version 3.0 site license for $500 (Cdn). A colleague from another university inquired whether the deal was open to universities. Apparently, it wasn't. Due to drastic cutbacks in Canadian provincial education budgets, no Canadian school likely could afford ArcView on a per-seat basis.

It makes economic sense to offer a discount deal to schools because there are so many and because they couldn't otherwise afford the software. Some universities, however, could acquire the necessary funding to purchase the software, so why offer them the same deal? Who cares if grade schools have better resources than university departments? It appears to be good economics, at least in the short term, but is it good for schools?

Pros and Cons

The merits of computers in schools have been debated for years, but recently the debate has intensified and the stakes have risen. One of the strongest endorsements for computers came from a report prepared by London-based McKinsey and Co. for the National Information Infrastructure Advisory Council. The report seeks to "help policy makers and educators understand what it will take to connect the nation's public K-12 schools to the information superhighway over the next five to 10 years." The full report can be accessed, appropriately enough, via the Web at *www.uark.edu/mckinsey/contents.html.*

The McKinsey report is well documented and enthusiastic about the new technology. In the executive summary, the report's authors state that access to the information highway "provides teachers access to a broad array of online curricular materials and innovative instructional approaches. It also opens up new communications channels between students and teachers, teachers and their peers, and schools and their communities."

On the other side of the argument is Todd Oppenheimer (1997), who noted the failed educational technologies of previous decades, ranging from Thomas Edison's motion pictures of the 1920's to B.F. Skinner's teaching machines of the late 1950s. Although computers, Internet access and related technologies received endorsements from the Clinton administration and Republican Newt Gingrich, Oppenheimer remains unimpressed.

"There is no good evidence that most uses of computers significantly improve teaching and learning, yet school districts are cutting programs—music, art and physical education—that enrich children's lives to make room for this dubious nostrum," he comments.

Oppenheimer cites advocates for low-technology approaches to teaching (Healy 1990;Postman 1992, 1996), but his observation, that one study suggests the Reader Rabbit program—used in more than 100,000 schools—caused a 50% drop in student creativity, isn't properly documented.

What About GIS?

Suppose for a moment we believe the critics. Suppose we feel an experiential approach to learning is more valid, that computers take funds away from more deserving needs, that there's a lack of courseware (as even the McKinsey report admits), that we're drowning in too much information (Postman 1996) and that an electronic culture is less preferable to a written one (Postman 1992). What would this mean for GIS in schools? Are these theories valid for GIS, and if not, when should we introduce GIS to students?

First, we never should sacrifice field trips in the physical or human environment for GIS or the "virtual field trip." Second, we never should reduce the number of quality of geography teachers to buy computers or GIS licenses. Third, although there's a lack of GIS courseware, the situation is improving. For example, Bob Sharpe and Wilfrid Laurier geography department, University of Waterloo, Ontario, Canada, presented at GIS '97 a GIS certificate program and model GIS curriculum for Ontario high schools. In addition, Sarah Bednarz (1995) described efforts of the National Center for Geographic Information and Analysis and National Council for Geographic Education to place GIS education in our schools. Also, Redlands, Calif.-based ESRI Inc. has produced tools to help grade-school teachers introduce GIS, including white papers—available on ESRI's Web site at *www.esri.com* — titled "GIS in K-12 Education," "Explore Your World" and "Introduction to ArcView for Schools and Libraries." Another useful resource is the conference report of the First National Conference on the Educational Application of GIS produced in 1994 by TERC Communications, Cambridge, Mass.

Finally, let us never privilege the electronic culture over others. The oral and written cultures have their merits, too.

(This column originally appeared in the November 1997 issue of GeoWorld.*)*

Section 5. References

Bednarz, S. 1995. "Reaching New Standards: GIS and K-12 Geography," *GIS/LIS '95 Proceedings*, pp. 44-52. American Society for Photogrammetry and Remote Sensing, Bethesda, MD.

Boserup, E. 1965. *The Conditions of Agricultural Growth*. Allen & Unwin, London.

Bradshaw, W.G., Webster, A.L., McRae, S.D. 1987. "GIS and Wildland Fire Prevention: An On-the-Ground Test." *Proceedings of GIS-87, San Francisco*. American Society of Photogrammetry and Remote Sensing, Bethesda, MD.

Brin, D. 1998. *The Transparent Society*. Perseus Books, Reading, Mass.

Burrough, P.A., McDonnell, R.A. 1998. *Principles of Geographical Information Systems.* Oxford University Press, Oxford, England.

Clarke, K.C. 2001. *Getting Started with Geographic Information Systems (Third Edition)*. Prentice Hall, Upper Saddle River, N.J.

Coates, J. 1992. "Computer Overlords." *Calgary Herald*. Aug. 29, 1992, p. D8.

Craig, W.J. 1993. "A GIS Code of Ethics: What Can We Learn from Other Organizations?" *Journal of the Urban and Regional Information Systems Association*, Vol. 5, No. 2, pp. 13-16.

Craig, W.J. 1993. "A GIS Code of Ethics: What We Can Learn from Other Organizations?" *Journal of the Urban and Regional Information Systems Association*, Vol. 5, No. 2, pp. 13-16.

Crampton, J. 1995. "The Ethics of GIS." *Cartography and Geographic Information Systems*, Vol. 22, No. 1, pp. 17-29.

Currie, M.J. 1995. "Rethinking Rights and Responsibilities in Geographic Information Systems: Beyond the Power of the Image," *Cartography and Geographic Information Systems*, Vol. 22, No. 1, 58-69.

De Mott Curtright, E. 2000. "Guide to PC Privacy: Create a Privacy Policy for Your Web Site," *Smart Computing*, Vol. 8, No. 4, pp. 131-133.

Dobson, J.E., Fisher, P., 2003. "Geoslavery," *IEEE Technology and Society Magazine*, Spring 2003, p. 47.

Dobson, J.E. 1993. "Consider Both Sides of GIS Ethics," *GIS WORLD*, Vol. 6, No. 9, pp. 20-21.

Dobson, J.E., 2000. "What Are the Ethical Limits of GIS?" *GeoWorld*, Vol. 13, No. 5, p. 24.

Durocher, J.M., 2002. "Webraska CEO Proposes 'Laws of LBS'"; *www.m-travel.com/20614.shtml*

Farwell, J. 2000. "Guide to PC Privacy: Online Profiling," *Smart Computing*, Vol. 8, No. 4, pp. 24-26.

Fisher, P.F, Dobson, J.E., 2003. "Who Knows Where You Are, and Who Should, in the Era of Mobile Geography?" *Geography*, Vol. 88, No. 4, pp. 331-337.

Foucault, M. 1979. *Discipline and Punish*. Vintage Books, Random House, New York.

Giddens, A. 1985. *The Nation-State and Violence*. Polity Press, Cambridge, UK.

Glass, G. V. 1976. Primary, Secondary, and Meta-analysis of Research. *Educational Researcher*, Vol. 5, pp. 3-8.

Gore, A. 1991. "Infrastructure for the Global Village." *Scientific American*, Vol. 265, No. 3, pp. 150-153.

Harvey, F. 2003. "The Significance of Public Safety for GIS Professional Licensing and Certification," *URISA Journal*, Vol. 15, No. 1, pp. 15-20.

Healy, J.M. 1990. *Endangered Minds: Why Our Children Don't Think*. Simon and Shuster, New York.

Kemp, K.K., Goodchild, M.F., Dodson, R.F. 1992. "Teaching GIS in Geography." *The Professional Geographer*, Vol. 44, No. 2, pp. 181-191.

Kleinfeld, J. 1999. "Student Performance: Males vs. Females." *The Public Interest*, No. 134, pp. 3-20.

Linden, D. 1991. "Comments from the Board: Ethics and the GIS Industry—Are the Terms Contradictory?" *GIS WORLD*, Vol. 4, No. 4, p. 17.

Lyon, D. 1994. *The Electronic Eye: The Rise of the Surveillance Society*. University of Minnesota Press, Minneapolis.

Mander, J. 1991. *In the Absence of the Sacred: The Failure of Technology and the Survival of the Indian Nations*. Sierra Club Books, San Francisco.

McPhee, J. 1989. *The Control of Nature*. Farrar, Straus & Giroux, New York.

Moravec, H. 1988. *Mind Children: The Future of Robot and Human Intelligence*. Harvard University Press, Cambridge, Mass.

Myrdal, G. 1957. *Economic Theory and Underdeveloped Regions*. Duckworth, London.

Norman, D. 1992. *Turn Signals Are the Facial Expressions of Automobiles*. Addison-Wesley, New York.

Obermeyer, N. 1998. "Professional Responsibility and Ethics in the Spatial Sciences," *Policy Issues in Modern Cartography*, Pergamon, New York.

Oppenheimer, T. 1997. "The Computer Delusion," *The Atlantic Monthly*, Vol. 280, No. 1, pp. 45-62.

Orwell, G. 1990. (first published 1949). *1984*, The New American Library Inc., New York.

Pagels, H.R. 1989. *The Dreams of Reason*. Bantam Books, New York.

Pickles, J. (Ed.) 1994. *Ground Truth: The Social Implications of Geographic Information Systems*, Guilford Press, New York.

Pirsig, R. M. 1975. *Zen and the Art of Motorcycle Maintenance*. Bantam Books, New York.

Poster, M. 1990. *The Mode of Information*. University of Chicago Press, Chicago.

Postman, N. 1992. *Technopoly: The Surrender of Culture to Technology*. Knopf, New York.

Postman, N. 1996. *The End of Education: Redefining the Value of School*. Vintage Books, New York.

Regis, E. 1990. *Great Mambo Chicken and the Transhuman Condition: Science Slightly Over the Edge*. Addison-Wesley, Reading, Mass.

Rheingold, H. 1995. *The Millennium Whole Earth Catalog: Access to Tools & Ideas for the Twenty-First Century.* The Point Foundation, Sausalito, Calif.

Schultz, T.P. 1990. "Introduction." in Boserup, E. (ed.) *Economic and Demographic Relationships in Development.* Johns Hopkins University Press, Baltimore, MD.

Snow, C.P. 1959. *The Two Cultures and the Scientific Revolution, The Rede Lecture for 1959*. Cambridge University Press, Cambridge.

Sommers, C.H. 2000. *The War Against Boys: How Misguided Feminism is Harming Our Young Men*. Simon & Schuster, New York.

Sweet, M. 2000. "Guide to PC Privacy: Protecting Privacy in a Digital World," *Smart Computing*, Vol. 8, No. 4, pp. 4-8.

Taaffe, E.J., Morrill, R.L., Gould, P. 1963. "Transport Expansion in Underdeveloped Countries: A Comparative Analysis." *Geographical Review*, Vol. 53, pp. 503-529.

Tenner, E. 1991. "Voice Mail and Fire Ants." *The New York Times*, July 26, 1991, p. A11.

UCGIS. 2003. "The Strawman Report," Task Force on the Development of Model

Undergraduate Curricula; *www.ucgis.org/priorities/education/strawmanreport.htm.*

Voyer, D., Voyer, S., Bryden, M.P. 1995. “Magnitude of Sex Differences in Spatial Abilities: A Meta-Analysis and Consideration of Critical Variables,” *Psychological Bulletin*, Vol. 117, pp. 250-270.

Waters, N., 1993. “Do You See What I Hear; Do You Hear What I See?” *GIS WORLD*, Vol. 7, No. 12, p. 68.

Weinberg, N. 1990. *Computers in the Information Society*. Westview Press, Boulder, Colo.

Section 6.
GIS, Fieldwork and Ubiquitous Computing

Introduction
By Matt Ball

With the advent of geographic information system (GIS), Global Positioning System (GPS) and remote-sensing technology, the art of mapmaking became a more "remote" pursuit. No longer was a direct connection with the environment needed to create accurate maps.

In Dobson's opening column in this section, he relates a field experience that got him out from behind the data he had been analyzing in the office, leading him to make a breakthrough in the field. The importance of fieldwork is emphasized, but rather than call to leave all technologies behind when conducting field research, Dobson calls for the use of field stations to greatly improve productivity.

Dobson relates the geographer's field tradition and the need for fieldwork, stating that "many real-world problems, both physical and cultural, cannot be solved without direct observation and engagement in the field."

The ability to take GIS to the field has greatly improved during the last few years. Improved hardware capabilities, increased GPS receiver accuracy, improved wireless communication, advancements in software operating systems designed specifically for mobile users and mobile-specific application software all play significant roles in this evolution. As devices with more capabilities come online, hopefully we'll all get "out of the office" more to enhance our data and interact meaningfully with the world around us.

Waters' discussion of the exercise that was done to put a value on ecosystem services is a fascinating use of GIS and global datasets for the broadest possible picture of our world. It's interesting to note that this survey excluded an assessment of the Antarctic. In the face of current global climate change and sea-level rise, the value of such an exercise seems all the more relevant, and a deep understanding of what we'll miss seems necessary.

Sui's exploration of the impact and implications of ubiquitous or pervasive computing on GIS is timely, as the number of hardware choices, communication platforms and sensing technologies continues to grow at a rapid pace. One can see that such rapid development with so many different options is leading to a saturation point where we'll always be connected.

The paper by Sui and Goodchild on "GIS as Media" seems particularly relevant in this constantly connected environment. This paper has resonated with me for some time, and I see its validation in the free services that the likes of Microsoft and Google provide. These providers give away content for free in exchange for the traffic that they sell to advertisers.

The evolution from media to the ambient intelligence that Sui describes, where "the user is surrounded by a multitude of interconnected, invisibly embedded computer systems" that "recognize users, adapt to their preferences, and offer natural ways of interaction" will undoubtedly lead to the "context-aware" computing that he foresees. The conflicting possible outcomes of this level of connectivity that he closes with leave much room for debate about the role of GIS in the "post-human age."

Fieldwork in a Digital World

By Jerry Dobson

In 1993 *Weekly World News* created a sensation at checkout lanes everywhere with the alarming announcement that a strange creature—half-man, half-alligator—had been discovered in a Florida swamp. Geographically, it turns out, the report left something to be desired. Four months later, an investigative report in Outside magazine improved the spatial accuracy and precision of the incredible discovery with a well-documented claim that the beast actually resided in Marsh's Free Museum in Long Beach, Wash. (White, 1994). Even after Outside's convincing report, a newspaper somewhere back East denied the existence of both the museum and the creature. It seems that a skeptical reporter had checked the phonebook for Long Beach, Calif. rather than Long Beach, Wash. Adding to the confusion, the museum's promotional postcard unequivocally states that "Jake the Alligator Man"—as he is affectionately called by those who know him best—actually came from "a famous California museum." But the eponymous Mr. Marsh himself knows a man who knows a man who saw a similar creature in a swamp somewhere in East Texas.

Now, with geographic information systems (GIS), the global positioning system (GPS), and field-station technology, field investigators need never again confuse Florida, Washington, California, and Texas again. The precise coordinates of Jake, the display case he calls home, the museum, and the town of Long Beach can be reported to a latitude/longitude precision of .01". Of great interest to me, though perhaps not to you, is the additional, derived information that Jake's waypoint lies on an azimuth of 76° precisely 3,479.614 kilometers from my own waypoint at that time in Oak Ridge, Tenn. Ed Bright, Don Field, and I discovered these amazing facts while conducting field verification of classified satellite imagery on the Washington coast.

I confess, however, that Jake has little to do with the rest of this essay. In the spirit of tabloid headlines at grocery stores near you, I shamelessly used him to get your attention. My real intention is to write about fieldwork and the remarkable improvements that GIS, GPS, and field-station technologies have brought. For what it's worth, however, we did detect stitch marks on Jake's torso that might suggest some sort of post-mortum, batch-mode gene splicing.

Digital Discovery

Once on a remote mountainside in upstate New York, I compared the landscape before me with a satellite image and suddenly realized that the forest was undergoing massive changes—changes affecting hundreds of square kilometers that, in theory, might be related to the landscape processes causing acidification of some Adirondack Mountain lakes (Dobson 1990). Without that synoptic view in my hand, I would have seen the trees but not the forest. The elation I felt at that initial moment of discovery could not have been greater if I'd found an uncharted island at sea, and the research that followed was as intriguing as any mystery could be. For the next three years my colleagues and I conducted research, examining the Adirondack case in detail and discovering similar cases in Sweden and elsewhere. We concluded that the suspected relationship was correct and could have profound implications regarding the impacts of acid deposition. The discovery and subsequent research were published in the *Annals of the Association of American Geographers* and one reviewer called it a "breakthrough" in the understanding of lake acidification (Dobson, et al. 1990).

That revelation was based on a page-size hardcopy image produced in the laboratory and carried in my backpack. I still recall how much I missed the heavy reams of paper maps left in my car and multiple satellite images stashed in the lab. Wouldn't it be great, I thought, to bring all those maps and images to the field in a portable computer? Better yet, wouldn't it be grand to have the computing power to process those images and maps, incorporate my new observations, and conduct geographic analyses in the field? That was in the mid-1980s, and I suspected the day would come when all those wondrous things would be possible.

Field-station Technology

In 1991 I bought my first hand-held GPS receiver, just before Allied forces cornered the world supply for Desert Storm. We paid $3,700 for a five-channel receiver, antenna and accessories. Today far better 12-channel models are available for $250 or less, and they're used routinely by hunters and boaters. Commercial software was slow to develop, but by 1993 we routinely conducted fieldwork digitally on laptop computers with satellite imagery, derived land-cover classifications, roads, and other key databases. With the attached GPS continually feeding coordinates to the screen, we were able to follow a "YOU ARE HERE" dot on the map. We couldn't actually conduct GIS analysis or image processing in the field then because laptop computers weren't powerful enough to support the applications software, but that is routinely possible today.

Conducting fieldwork in Maine and New Brunswick in October 1993, Ed Bright and I spent three days rushing from site to site in a rugged fall landscape more beautiful than I can describe. Our purpose was to verify his classification of Landsat Thematic Mapper imagery for the Coastal Change Analysis Project (C-CAP) (Dobson and others 1995). When detecting change, classification accuracy and precision are crucial because errors in either time period show up as false change. The GPS made a difference, but the biggest advance came with the combined field station that related GPS coordinates to satellite imagery and other databases in a GIS framework.

The components of my first field station included an IBM ThinkPad (Intel 486, 50 Mhz, 16 MB RAM, 160 MB hard disk), Magellan NAV 5000 PRO GPS, FieldNotes software, paper maps, rental car, what seemed like 23 miles of tangled cable, and a cardboard box (Dobson 1994a). Our Oak Ridge National Laboratory colleague, Rick Goeltz, surveyed the options and assembled the system. All the components were commercially available and all of them actually worked, though each was disappointing in its own way. Before leaving Oak Ridge, Ed Bright loaded nearly an entire Landsat scene in raw and classified form. He also registered the road network from the Census Bureau's Topologically Integrated Geographic Encoding and Referencing (TIGER) files to the same map projection and incorporated it as a separate layer. FieldNotes accepted coordinates live from the GPS and placed them on the screen, along with raster images and vector databases. The result dramatically improved productivity. On our very first field trip, we covered two or three times as many sites per day as we ever had, and every site counted.

The new technologies were remarkable, but it was obvious that the designers hadn't tried their components under actual field conditions. The cardboard box was our hasty solution, for instance, to the near fatal flaw that passive matrix color screens turned as dark as night when exposed to the light of day. We called it the "Shroud of Tourin'" and recommended a cardboard box with three holes, one cut high for viewing and two cut low for access to the keyboard. We never found a solution to the tangle of cables, though we suspected that the GPS and laptop manufacturers could have done so if they'd tried. We bought a commercial backpack specifically designed for carrying a laptop and GPS but soon discarded it. We found it more of a burden than the equipment carried alone. Its design was like a laundry bag with excessive bundles of belts and buckles attached. An external frame pack would have made a better choice, offering more separate pouches. With modifications, its frame conceivably could have been turned into an antenna.

That first trial was in Maine and New Brunswick in the fall of 1993. The following spring we returned, and in June 1994 we headed for Oregon and Washington, working the coast from Tillamook Bay to Willapa Bay and encountering Jake along the way. Later we used the same setup in North Slope and Southeast Alaska. Our field-station configuration was unchanged except for great improvements in the Shroud of Tourin' (Dobson 1994b). Gone was the old, battered cardboard box of Version 1.0. The new Version 2.0, based on advanced coat hanger and black felt technology, opened to rave reviews on both coasts. Soon, active matrix screens rendered the Shroud itself obsolete, but I saved one for old time's sake.

Good field-station software supports field observation though automated field forms that can be called to the screen at any time or place. They can be customized with menus and queries that prompt specific observations (vegetation, air quality, ownership, religion), or they can record free-form comments. In either case, geocoordinates are entered automatically with each observation, and the completed form is entered automatically into a spreadsheet or database management system. We've recorded as many as 300 separate observations per day from cars, airplanes, helicopters, boats, all-terrain vehicles and airboats.

Unfortunately, laptops still aren't practical for the type of wilderness fieldwork that I did so extensively in the Adirondack Mountains. Hiking on foot with a backpack, the weight of a laptop alone would be a burden, but the greatest problem is batteries.

Laptop manufacturers live by The Big Lie, and the biggest lie of all is battery life. Most batteries don't last more than, say, two and a half hours, and they're heavy. Take enough of them to last a few days, and you'll have to leave your tent at home. And they're expensive. You can buy one or two of them for the money you save on the tent.

All of my experience with the field station has focused on physical topics, but I can conceive of many cultural applications as well. I currently direct the LandScan Global Population project that annually produces a global population database at 30 arc sec (1 km or finer) resolution for the entire world (Dobson and others 2000). We'd like to verify our results in the field, but we haven't because, short of door-to-door counts, in situ estimates of the number of people per cell would be subjective and unreliable. Now, we're funded to produce a far more precise (3 to 5 arc sec resolution) database for the United States, and we expect to sample many areas with field station in tow.

It will be a bigger leap, of course, for cultural geographers who need to interact with people in their homes and communities. Technically, there's no reason the field station can't serve them in the same way, but they'll have to judge whether the presence of technology per se inhibits or otherwise influences their subjects.

In C-CAP we often needed to interview local residents to learn the history of land cover change on particular plots of land. In rare instances the field station sparked suspicion and interviewees clammed up. More often, after a brief explanation, seeing their personal geographical space onscreen aroused curiosity among interviewees and led them to be more open, sometimes effusively so. The synoptic view often led residents to respond more geographically, sometimes with revelations of their own. I recall, for example, one man exclaiming, "I didn't know there was a lake back there," as he spotted a patch of blue pixels behind a hill beside a road he'd driven all his life. One can easily imagine similar cultural revelations—a church, a community, a noxious facility whose prominence, prevalence, or proximity might surprise local residents. Such revelations may help in certain types of cultural investigation. Conversely, they may bias results in other types. That will be for investigators to decide. At a minimum, it should be useful to enter information from interviews and other field observations—cultural or physical—into automated field forms with GPS coordinates automatically attached.

Handheld computers, such as Palm, offer reasonable alternatives that may be less intimidating for cultural fieldwork and certainly would be lighter and longer-lived for wilderness fieldwork. Screen size, storage capacity, and processing power are limited at present, but manufacturers are making great strides. Wireless communication through the Internet or an organization's own intranet can alleviate some of the storage problem. Better yet, Internet Mapping and Internet-GIS software increasingly support map viewing, and even data manipulation, at a distance.

Field-testing the Mazda 626

Motorized transport is essential for long distances, of course, and even for short distances when bearing electronic equipment. With digital fieldwork there's a new concern for elapsed time. Even when steadfastly ensconced in a single place, you'll need a power supply for recharging batteries, and vehicles usually perform that function well.

Always ask yourself just how important it is to go off road, on the water, or up in the air considering how much time and cost are incurred with alternative vehicles. We've used airplanes in Southeast Alaska, seaplanes in the Adirondacks of New York, helicopters on the North Slope of Alaska, boats on the Columbia River of Oregon and Washington, all terrain vehicles on the coast of Maine, and airboats in the Everglades of Florida. Every one of them presents special problems ranging from electrical supply (cigarette lighters, receptacles) to mounts (convenient and safe places to stash equipment) to visibility. In the airboats, for example, saw grass continuously thrashed our faces and the laptop screen, and seeds turned up in our keyboards for weeks afterward.

Fieldwork often means going places that can only be reached by all-terrain vehicle or rental car. Even when sport utility vehicles are available, the ubiquitous sedans offer advantages of cost, availability, and versatility. As a result of one trip to the salt flats of Utah several years ago, I still hold the World Land Speed Record for a Rented Malibu Classic. Often the only available option is a rental car, and that can be the weakest link in the system. Field testing the Mazda 626, we found it had the style and performance for all terrain but less than optimal clearance. Even so, the car sustained only minor damage, mostly cosmetic, to it's aptly named rocker panel. On another trip we found the Olds Cutlass Ciera less satisfactory than other models we'd used in the past. The headliner was inadequately padded, and the front cross member of the frame was too malleable. Once, when we struck a boulder, Ed Bright hit the rof with such force that he bloodied his noggin. I thought the lights had gone out, but it turned out my cap had just fallen down over my face. Also, it seems General Motors' designers had mistakenly assumed that all vertical forces would be downward when they designed the motor mounts, and that error was further compounded by placing entirely too much breakable hardware between the motor and the hood. Ultimately, we pledged never to undertake any fieldwork worth less to society than the better part of an Olds Cutlass Ciera, and the pledge has not been difficult to keep. Recently, it was announced that General Motors had reached a similar conclusion and would discontinue the entire Oldsmobile line.

Foreign Fieldwork

Foreign study raises additional issues of technical compatibility, data availability, physical security, and political acceptance. So far, I haven't experienced any political or technical difficulties with the laptop, even in countries that distrust Americans. Twice, however, we were questioned at the Canadian border, of all places, once over an egg-shaped GPS antenna on our vehicle's roof and once over remote-sensing hardware in our baggage. Once, in Norway, I was unable to connect via modem despite having consulted local experts and having bought a new adapter. Later, I bought a kit of 42 electrical and telephone adapters that's supposed to cover the entire world, but it doesn't, and I sometimes have to buy adapters in-country anyway.

I've carried my laptop to many foreign countries, and it's often a judgment call. A trip to Eritrea in the summer of 2001 illustrates a common dilemma. The country has been at war for 30 years, with border fighting as recently as 1997. Some reports say it's peaceful now, but some of those same reports say the train still runs from Asmara to Massawa, and that hasn't been true for decades. A friend told us that one of his colleagues was fired on not long ago as his plane landed in Asmara, and United Nations peacekeepers are prominent throughout the country. Considering the uncertainty, I opted to leave my laptop and GPS receiver behind. It turned out that neither item would have caused a problem. Laptops must be registered on entry, but there's no difficulty taking them back out. When I hired a boat to explore the Dahlak Banks, the proprietor was surprised and disappointed that I hadn't brought a GPS receiver along.

Still, this story illustrates how believable my concern was at the time: Professor Emeritus Bob Peplies of East Tennessee State University traveled with me to Eritrea. At midweek I headed down to the Red Sea with Habte Frezghy, who frequently served as our driver. (In many foreign countries, it's best to hire a car and a driver.) We agreed that Bob would stay in Asmara and ride down to meet me four days later. Descending for three hours through rugged, arid terrain, I was struck by a landscape feature that I'd never seen anywhere else. I called Bob to tell him about it. "Bob, you've got to see this," I said. "This whole country is terraced. Every hill, every mountain. . . bottom to top. It must have been going on for ages." It's done purely to hold enough soil and moisture to grow trees.

Four days later Bob headed down the mountain looking behind every one of those trees. Only when he saw the terraces himself did he realize that I had said "terraced." My hat's off to Bob, who was brave enough to take that ride, thinking I'd said "terrorist."

Field Navigation

No doubt, some of you take pride in your orienteering skills and may not consider GPS or field stations necessary. But consider the case of my partner, Ed Bright. Ed is at one with the universe. He doesn't need a GPS because he always knows where he is. Even when the printed map is wrong or out of date, Ed knows he's right and the map is wrong. (I'm not being facetious; he is right.) Once, as we drove across New Brunswick, I challenged Ed to guess the distance to London, England. He placed his hand on his head for no more than 20 seconds and said, "3,000 miles." I entered our waypoint into the GPS, and the answer came back, "3,003 miles." (Yes, this actually happened.) Ed knew our coordinates precisely without looking at the GPS. He knew that our location was approximately on the same latitude as London, and he took Greenwich, England's 0° longitude as a surrogate for London. He mentally calculated the length of a degree of longitude, using the cosign of our latitude, and multiplied that by the difference in longitude. Still, for Ed and others with special orienteering skills, the GPS advantage is that they spend less time arguing with those of us who are occasionally at odds with the Universe.

Anyone who has tried to check land-cover classifications derived from aerial photography or satellite imagery knows that the overriding problem is spatial orientation—knowing where you are on the ground and on the map, image, or database. A careful analyst may reject several sites per day due to uncertain location. With only a paper map, orientation depends on spatial pattern recognition, a very demanding form of reasoning that drains energy and occupies time. The field station displaces this exhausting activity and allows workers to concentrate on understanding the landscape.

"Real-time" GIS

"Real-time" GIS emerged as a prominent new item when the University Consortium for Geographic Information Science (UCGIS) called for nominations of research priorities in 1996 (Dobson 1996). Laptop computers, GPS, high-speed wireless information links, fast multimedia satellite transmission, high-performance computing, and content-based retrieval from digital libraries have opened a whole new realm of information support in the field. So have improved GIS software, the Internet, and CD-ROMs. I've barely begun to tap the new potential, and already it has transformed my access to information, especially on remote travel, when I would have been utterly disconnected before. Let me illustrate with a trip to one of my favorite destinations, Yakutat, Alaska, taken in 1996.

My sole purpose in recounting this trip is to convey a sense of what it's like to travel in an information-rich fashion. All the events actually happened, although I've compressed them from several trips into one, and all of the information activities have been undertaken, although perhaps not in the stated circumstances. Still, this fanciful trip was just a taste of what's possible, and the capabilities are growing daily.

We took Alaska Airlines from Juneau to Yakutat, where a reservation had been made for a flight over the Hubbard Glacier. Just before taking off in Juneau, we learned that our bush pilot had been killed in a crash the day before. We shared the flight with ice climbers on their way to retrieve his body from a mountain he failed to see because of clouds. When the circumstances were explained, it was clear that GPS with an onscreen aeronautical chart could have prevented the accident.

We landed in Yakutat and taxied toward the terminal. A huge sign over the Yakutat Lodge said, "FOOD, SHELTER, BOOZE"—everything a traveler really needs right there in one building.

On the previous trip we had been invited to dinner with the owners of a bed-and-breakfast on the outskirts of town. This time, all I could recall was that the inn's name was "Blue" something or other. I popped in my PhoneDisc Powerfinder CD-ROM and searched for all names beginning with BLUE in Yakutat, Alaska, and there it was:

BLUE HERON INN BED & BREAKFAST
Box #254
YAKUTAT, AK 99689-0254
907-784-3287

I didn't want to admit that I'd forgotten the owners' names, so I copied the listing, switched to a search by telephone number, and presto: "John and Fran Latham."

I couldn't recall precisely how to find their place, so I inserted DeLorme's Street Atlas USA. Ordinarily I would have entered a street address and zip code, and the software would have displayed a detailed street map with the house on the correct side of the street very close to its actual location. E- 911 has ensured that street names and numbers are available for most cities and many rural areas but not yet for Yakutat at that time. We liked the town because it(s in the wilderness, so we could hardly complain that it didn't have street numbers.

We rented a Chevy Suburban at Situk Leasing and rigged it for GPS linked directly into our field station. We inserted a CD-ROM prepared by the National Oceanic and Atmospheric Administration that contained Ed Bright's Thematic Mapper satellite classification for the scene covering Yakutat, Russell Fiord, and the Hubbard Glacier. We zoomed in on the Blue Heron's neighborhood and figured out the location based on our recollections of surrounding land cover. The field station ticked off a continuous track of GPS coordinates as we made our way to the inn.

Even in 1996, I was impressed with the selection of U.S. atlases available, but I was disappointed with the world atlases that were around back then. Their prevailing perspective seemed to be entertainment and visualization, rather than analysis. Still, in a phone conversation from Yakutat, when my brother and book coauthor, Jeff Dobson, asked the burning question, "Is it possible to see Mount Olympus from the Acropolis in Athens, Greece?" I was able to answer in less than a minute using ABC's 3D Atlas. Later I bought Microsoft's Encarta World Atlas, and it's been my traveling companion ever since.

Practical Advice

I'm still every bit as enthusiastic about the field-station technology as Ed and I were in the beginning, and we've learned some lessons that may help others. So here is some practical advice:

- Don't attempt to conduct field verification of remote sensing or detailed cartographic data without a GPS field station. The combination of GPS, raster images, and vector overlays is a quantum leap in the technology that supports fieldwork. It's difficult to convey in words just how effective the field station is. Once you've tried it, you'll never want to use hardcopy (paper or mylar) plots again.
- Don't assume anything will work unless you have tried it before departure. This goes for hardware and software, especially for connections of all types. On two occasions when we tried to assemble a second unit in the field using components that arrived with participants from different places, the unit didn't work.
- If you plan to use an all-terrain vehicle or rental car, buy a DC/AC converter for your laptop and a dual splitter for your cigarette lighter. These inexpensive devices will allow you to power both the computer and the GPS unit without depending on batteries.
- Treat every hardware component as fragile even if the sales hype says it's intended for the field. So far, we've had to replace the hinges on a laptop and the serial connector on a GPS receiver. Some laptops have far too many tiny doors that tend to bounce off and get lost.
- Don't be timid about customizing weak points in the system. On our first GPS receiver, for example, I cursed the flimsy wire that ran from the DC plug to the transformer and replaced it with a 16-gauge wire. I also soldered fork connectors at the transformer end of the new, heavier wire.

• Expect bad weather. Electronic devices are notoriously sensitive to temperature, moisture, and shock. Ideally, one would pre-test all equipment under the same extreme conditions expected in the field, but that won't be possible. The best one can do is think ahead and prepare to adapt to a variety of conditions. We tried "ruggedized" (the proper term for a laptop that's weather- and shock-resistant, not to be confused with "tempest hardened," which means it's shielded to prevent transmission of electromagnet pulse) "luggables" (a laptop that's too heavy and cumbersome for a standard lap), but we switched back to regular laptops as soon as possible.

• Load road networks and register them to your images so you can find your way to your study sites. Since 1990 TIGER data have worked exceptionally well for field areas in the United States. Today many other options, such as DeLorme's Street Atlas USA, are available on CD-ROM, and similar databases are slowly becoming available for other parts of the world.

• Bring along some nice, old-fashioned paper plots just in case the electronics fail or you catch some fish.

The Field Tradition

Geographers sometimes feel they were born too late, after everything had already been discovered. Many of us would like to be in terra incognita, adventuring and discovering places never seen before. Instead, we try to learn more about familiar places and the processes occurring in them. To us, this type of discovery can be as intriguing as a new island, river, or canyon might have been in earlier times, though it's difficult to generate the same level of excitement among the public. We continue to seek the field because we enjoy it and believe in it. We believe, in fact, that many real-world problems, both physical and cultural, cannot be solved without direct observation and engagement in the field.

The field tradition in geography used to be so strong that those who did not practice it were stigmatized as "armchair geographers" (Platt 1959). In the past decade GIS has increased the quantity and quality of field-based data in geographical analysis, but, at the same time, has enticed analysts to sit at the console while others collect the data. Recently GPS has taken field data collection to new levels of accuracy and precision. Now field stations, combining GPS and GIS technology in powerful, color laptop computers, are supporting the field tradition as never before. In the next phase geographic analysis, as well as data collection, will be done in the field.

During the Age of Exploration it was a stunning accomplishment simply to observe a new landmass and report it's existence. For several centuries thereafter, exploration meant filling the gaps in continental maps and reaching toward the Poles. Today, exploration is more demanding than ever as we probe deeper (oceans, inner Earth) and higher (atmosphere, planetary bodies), search for tangible evidence of historic and pre-historic landscapes, and seek new geographic understanding of complex earth processes. Through technology, exploration can now be extended to places where humans cannot go, to phenomena that cannot be observed directly by human senses, and to macroscopic processes so large they can be observed only through remote sensing, GIS, and three-dimensional geovisualization. Now, information technology opens new possibilities for involving the public in exploration. It would be technically feasible, for example, to provide near-real-time access to space imagery or ocean-floor imagery to anyone connected to the Internet or even to let private citizens actively direct sensors from their home computers.

In June 2000, the American Geographical Society council voted to revitalize the society's exploration program and seek to endow an exploration fund. Our goals are:

- To reestablish exploration as a key mission of the Society
- To promote exploration by government, business, and academia
- To lead and encourage the discovery of new knowledge about the earth and planetary bodies
- To encourage greater private participation in exploration of all types

All four goals, especially the final one, have been revolutionized by emerging information technologies that can now be deployed in the field and monitored or controlled over the Internet.

(This column originally appeared in the January/April 2001 issue of Geographical Review. *It is reprinted here with permission.)*

GPS Advances the Field Tradition in Geography

By Jerry Dobson

I am writing this manuscript while conducting fieldwork in New Brunswick, Canada and Maine, USA. The column won't be published until January, but it is now October and we have spent three days rushing from site to site in a rugged Fall landscape more beautiful than I can describe. My partner and I are verifying his classification of Landsat Thematic Mapper imagery for the CoastWatch Change Analysis Project (C-CAP). In change detection, classification accuracy and precision are crucial because errors in either time period show up as false change. We have done this many times in the Chesapeake Bay, the Adirondack Mountains, Alaska, Washington, Oregon, and elsewhere, but we could not have done it this way before. Global Positioning Systems (GPS) made a difference, but the biggest advance came with fieldstations that relate GPS coordinates to satellite imagery and other databases in a GIS framework.

No doubt, some of you take pride in your ability to navigate and may not consider GPS or fieldstations necessary. But consider the case of my partner, Ed Bright. Ed is at one with the Universe. He doesn't need a GPS because he always knows where he is. Even when the printed map is wrong or out of date, Ed knows where he is. Just yesterday as we were driving across New Brunswick, Ed guessed the distance to London, England with an error of only 3 miles in 3,000. Still, for him the main advantage of GPS is that he has to spend less time arguing with those of us who are occasionally at odds with the Universe.

Anyone who has tried to check land cover classifications derived from aerial photography or satellite imagery knows the overriding problem is spatial orientation—knowing where you are on the ground and on the map, image, or database. A careful analyst may reject several sites per day due to uncertain location. With only a paper map, orientation depends on spatial pattern recognition, a very demanding form of reasoning that drains energy and occupies time. The fieldstation displaces this exhausting activity and allows workers to concentrate on understanding the landscape. The best testimony to this improvement is the fact that I feel like writing each evening after as long as 12 hours in the field.

Fieldstation Technology

The fieldstation components include an IBM Thinkpad (486, 50 Mhz, 16 Mb RAM, 160 Mb hard disk), Magellan NAV 5000 PRO GPS, Field Notes software, paper maps, rental car, 23 miles of tangled cable, and a cardboard box (Figure 1). Our coworker, Rick Goeltz, surveyed the options and assembled the system. All the components are commercially available and all of them actually work, though none are perfect. Before leaving Oak Ridge, Ed loaded nearly an entire Landsat scene in raw and classified form. He also registered the road network from Census's TIGER files to the same map projection and incorporated it as a separate layer for the U.S. portion of the scene. Field Notes accepts coordinates live from the GPS and places them on the screen along with raster images and vector databases. Imagine a symbol, like the "YOU ARE HERE" arrow on a wall map, following your every move. The result dramatically improves productivity. On this trip, we covered many more sites per day, and every site counted.

The new technologies are remarkable, but we have a feeling the designers of each component haven't tried their products under actual field conditions. The cardboard box is our hasty solution, for instance, to the near fatal flaw that active matrix color screens turn as dark as night when exposed to the light of day. We call it the Shroud of Tourin' and recommend a Solo Plastic Cup box with three holes, one cut high for viewing and two cut low for access to the keyboard. We haven't found a solution to the tangle of cable, though we suspect the GPS and laptop manufacturers could if they tried. We didn't take our GPS backpack along because, on previous trips, we found it more of a burden than the equipment carried alone. Its design is very much like a laundry bag with excessive bundles of belts and buckles attached. An external frame pack would make a better choice, offering more separate pouches, and with modifications the frame conceivably could be turned into an antenna.

Fieldwork means going places you can only go in a four-wheel drive or a rental car. As a result of one trip to the salt flats of Utah several years ago, I still hold the World Land Speed Record for a Rented Malibu Classic. On our current trip we found the Olds Cutlass Ciera to be less satisfactory than other models we've used in the past. The headliner is inadequately padded, and the front crossmember of the frame is too malleable. Often the only available option is a rental car, and that can be the weakest link in the system. Our solution is a solemn vow not to engage in any enterprise worth less to public welfare than the better part of an Olds Cutlass Ciera.

The Field Tradition

Geographers sometimes feel they were born too late, after everything was already discovered. Many of us would still like to be in *terra incognita* adventuring and discovering places never seen before. Instead, we try to learn more about familiar places and the processes occurring in them. Often to us this type of discovery is as intriguing as a new island, river, or canyon might have been, though it is difficult to generate the same level of excitement among the public. We continue to seek the field because we enjoy it and believe in it. We believe, in fact, that many real world problems, both physical and cultural, cannot be solved without direct observation and involvement in the field.

The field tradition in geography (Platt 1959) used to be so strong that those who did not practice it were stigmatized as "armchair geographers." In the past decade GIS has increased the quantity and quality of field-based data in geographic analysis, but, at the same time, has enticed analysts to sit at the console while others collect the data. Recently GPS has taken field data collection to a new level of accuracy and precision. Now fieldstations, combining GPS and GIS technology in powerful, color laptop computers, are supporting the field tradition as never before possible. In the next phase geographic analysis, as well as data collection, will be done in the field.

(This column originally appeared in the July 1994 issue of GIS World.*)*

Global GIS Demands Global Fieldwork

By Jerry Dobson

My goal is to see the whole world before I die, and I still have a long way to go (on both counts, I hope). Recently, an acquaintance asked what I had learned from my world travels. I responded with heartfelt concerns about rampant population growth, regional conflicts and stresses on coastal ecosystems.

Afterward, however, I couldn't get the question out of my mind: What, indeed, have I learned from world travel that I might not have learned otherwise from books, videos or GIS data?

Safer than Taxis

I've been traveling to "the far corners" to update and refine estimates of populations at risk. In the last three years, I visited the Horn of Africa, northwestern South America, the Middle East and the Balkan Region. Next comes the Far East.

We're focusing first on populations at risk due mainly to regional conflicts. Then we'll focus on natural hazards, and we eventually plan to cover the whole world.

The early going, naturally, had caused more than one person to ask, "Aren't you afraid?" I responded with two comforting gems gleaned from global travel: 1) no matter how much they hate Americans, they always hate their neighbors more, and 2) even in the most dangerous places, the risk of hostile attack never exceeds the risk of an ordinary ride in an ordinary taxi.

For a geographer, travel is fieldwork ... and more. In a real sense, the world is our laboratory, and "lab time" is as essential for us as for any chemist or microbiologist. Fieldwork seems natural when the study area is small. No one should claim to be an expert on the geography of a particular city, for example, without spending a lot of time there. The same goes for countries, regions, continents and the whole globe.

Watching the news today, I can visualize the streets of Monrovia, Liberia, because I've been there. The true value of that long-ago fieldwork, however, extends far beyond visualization.

It's one thing to know, as even talk show hosts do, that Liberia was settled by freed slaves from America. It's quite another to experience, as I did in 1982, how thoroughly their descendants, called Americo-Liberians, still view themselves as an outpost of America. The U.S. dollar was their currency, the U.S. Constitution was their constitution, the U.S. flag (with nine stripes instead of 13) was their flag, and one of their nine counties was named Maryland.

Once I watched the bewilderment and disappointment of Liberian colleagues as they searched an American newspaper in vain for mention of President Reagan's deliberations with a Liberian official who was known to be visiting the White House that day. They expected a front-page article, and there was none at all.

Balcony Bushes

Sometimes insight comes in a single, mind-blowing scene. There was in 1982 a multistory building in downtown Monrovia whose plumbing had sprung a leak somewhere above the third floor. A sizable stream flowed from the town's water supply through an apartment, over a balcony and down to the sidewalk below. Pedestrians routinely darted behind the waterfall, and the effect was rather pleasant, like hiking beneath a natural one. But this waterfall was the sure sign of a failing nation, for on that balcony bushes 10 feet tall had grown from its flowing waters.

Sometimes insight comes in a story or even a single phrase. In the Balkans this summer, we met a retired British Army officer who had married a Bosnian woman. Bubbling with enthusiasm for his new bride, he tried to convey to us what was so appealing about her, and finally he blurted, "She has a penchant for *violence*." He said that as if violence was a good thing, because it has become so much a part of the culture.

Once on the streets of Sarajevo, Bosnia and Herzegovina, I asked Matt Dunbar, a University of Kansas graduate student who accompanied me, "What percentage of the buildings that face the mountains have bullet holes or worse?" His estimate, the same as mine, was nearly 100 percent (see Plate 11). That damage, the burned National Library, the "balkanization" of ethnic communities within the city, and hatreds expressed in ordinary conversations signify a most uneasy truce that could erupt anytime with little provocation.

Full Field

None of these insights can be found on maps or seen on overhead imagery. Most are difficult to capture in a database. Geographic analysis—at any scale, resolution or coverage—requires fieldwork, and GIS is no exception.

Before laptops and personal digital assistants, GIS gained the reputation of being a "newfangled," electronic version of "armchair geography." Even then, however, enormous amounts of fieldwork went into most GIS databases, and respectable analysts insisted on fieldwork.

Now, especially with the explosion of Global Positioning System technology, GIS is recognized as a means of collecting data in the field, with growing potential to support full-blown analysis in the field. The ideal, of course, is seamless analysis from field to laboratory and back again.

(This column originally appeared in the September 2003 issue of GeoWorld.*)*

GIS and the Value of Everything

By Nigel Waters

Oscar Wilde wrote that a cynic is "a man who knows the price of everything and the value of nothing." Several environmental scientists, economists and even geographers have attempted to put a price tag on the planet's ecosystem goods and services (Costanza 1997). As Pimm (1997) noted, these academics are no cynics themselves. They realize that the true value of ecological life-support systems is, in one sense, infinite. Simply put, humanity wouldn't survive without them. However, the scientists also believe that it's instructive to list the replacement cost of the ecological system services that are vital to maintaining and supporting Earth's flora and fauna. How did they attempt such a Herculean effort?

Putting a Price on Natural Systems

Examining how these scientists went about their work provides GIS professionals a methodology for pricing the natural systems that might be damaged by building a highway through a national park, for example. This would be vital information in any cost-benefit analysis.

The scientists' first step involved determining which of the main ecosystem goods and services they would evaluate (Daily 1997). They identified 17 major categories, including gas, climate and water regulation; disturbance regulation (e.g., ameliorating the effects of floods and other extreme environmental events); water supply; erosion control and sediment retention; soil formation; nutrient cycling; waste treatment; pollination; biological controls (e.g., prey/predator dynamics); habitat refugia for transient and other populations; food and raw material production; genetic resources; recreation resources; and cultural resources (the aesthetic, artistic, educational, spiritual and scientific value of ecosystems).

Sixteen primary biomes or ecosystem types were identified. These were then split into marine and terrestrial ecosystems. In turn, the marine systems were divided into open ocean and four coastal categories: estuaries, sea grass/algae beds, coral reefs and continental shelf areas. The terrestrial systems consisted of two forest systems (tropical and temperate/boreal), grass or rangelands, two types of wetlands (tidal marsh/mangroves and swamps/floodplains), lakes/rivers, desert, tundra, ice/rock, cropland and urban areas.

Relying on extensive previous research, Costanza and his co-authors determined a value for each ecosystem service/biome combination. This figure was expressed in U.S. dollars per hectare per year. The only task that remained was to multiply the value per hectare by the number of hectares. The dataset table in spreadsheet format and copious methodology notes may be downloaded from *Nature* journal's Web site at *www.nature.com* (users must register to gain access to the content).

Paul Sutton, one of the article's co-authors and a geographer from the National Center for Geographic Information and Analysis at the University of California at Santa Barbara, informed me he used a GIS to produce the article's world map of ecosystem services. GIS also might be used to determine the area of each biome, although it wasn't used in this study.

Sutton told me even more accurate estimates might be made if NASA's land-cover dataset were used. This dataset is being developed as part of the International Geosphere/Biosphere Program. It appears that future ecosystem valuations will rely more on GIS datasets and analysis.

Why Did They Do It?

The authors note that this type of exercise, although fraught with difficulties, helps to establish upper and lower limits on ecosystems' value. These limits were determined to be US$54 trillion and US$16 trillion per annum, respectively. In addition, the exercise assesses the relative magnitude of ecosystem services, which—if a middle range estimate of US$33 trillion is used—are about 1.8 times the current global Gross National Product. The research establishes a framework for future studies of this kind, just as Costanza and his colleagues built on the work of Daily (1997) and Pearce (1993), among others. Finally, the study shows where more work is needed and is provocative enough to stimulate further research and debate.

One of the main analysis problems was that the database used included no data for the desert, tundra and ice/rock biomes. Such huge areas as the Antarctic presumably are yet to be included in the analysis. Perhaps ongoing research conducted at the University of Calgary will provide better models of Antarctica's role in providing various ecosystem services (Giovinetto 1990). In addition, Costanza and his colleagues believe that more ecosystem services should be considered and that more realistic representation of ecosystem dynamics and interdependencies will help to increase the accuracy of their estimates.

A Bargain at Any Price

Ecologists and economists now pay more attention to the worth of ecosystem services (see Plate 12). They realize, to quote Wilde again, that "no man is rich enough to buy back his past." Once destroyed, the services that ecosystems provided may become priceless.

(This column originally appeared in the September 1997 issue of GIS World.*)*

Will Ubicomp Make GIS Disappear?

By Daniel Sui

While speculating about the future of computers in the 21st century, Mark Weiser argued back in 1991 that "the most profound technologies are those that disappear. They weave themselves into the fabric of everyday life until they are indistinguishable from it (p. 94.)." Although the computer as we know it has not vanished as Weiser predicted, recent advances in ubiquitous computing (ubicomp) have accelerated the pace toward the disappearance of computers, as more and more embedded computers are found in mobile phones, car navigational systems, gas pumps, ATM machines, electronic road/bridge tolls, precision agricultural products, retail point-of-sale systems, etc. (Mann, 1996; The U.S. National Research Council, 2001).

Additionally, radio chips, led by the radio frequency identification (RFID) technology, are designed to replace barcodes on manufactured objects. RFID, along with existing location technologies as reviewed by (Hightower & Borriello, 2001) will be able to make the location of every single entity on earth trackable. Hand-held communication media such as cell phones, PDAs, or iPODs can easily mutate into wearable remote-control devices for the physical world.

These embedded computers, though invisible to users, are fast approaching the power and complexity of desktop PCs (The U.S. National Research Council, 2001). According to an estimate by Bill Gates (2003) a typical middle-class American already interacts with 150 embedded systems every day, most of the time without knowing it. With the maturity of ubiquitous computing, these embedded computers—which use up to 90 percent of the microprocessors produced today—will inevitably perform more PC-like functions.

More importantly, advances in wireless networks will make these embedded computers communicate seamlessly with their traditional PC counterparts. According to the Semiconductor Industry Association (*www.sia-online.org/pre_statistics.cfm*) the world microchip industry will be producing one billion transistors per person on earth by 2010. Computing is becoming ubiquitous, at least in an increasing number of places in the developed world.

The Pervasive Third Wave

Ubicomp is also known as pervasive computing (Satyanarayanan, 2001), typically invisible and everywhere accessible within a particular locale. Ubicomp represents the third wave of computing according to Weiser (1993). The first wave of computing started with the invention of ENIAC in 1946, and ended with the appearance of the first personal computer in 1983. This period was dominated by mainframe computers, with one computer serving multiple users. The second wave of computing, from 1983 to present, has been characterized by personal computers with one person and one computer in an uneasy symbiosis, staring at each other across the desktop without really inhabiting each other's worlds. The third wave of computing, argues Weiser (1993), will be dominated by ubiquitous computing.

Unlike the previous two waves, ubicomp will operate in a mode where many computers serve one individual, regardless of where the person is located in the world. Instead of interacting across the desktop, computers will be embedded into human bodies, furniture, walls, buildings, surrounding environments, neighborhoods, and cities, thus forming a seamless digital habitat. Ubicomp is heralding a new age that is full of smart mobs and sentient things (Rheingold, 2002). Indeed, the world predicted in David Brin's (1990) science fiction novel *Earth*, in which everybody is connected with everybody and everything else, has arrived much sooner than anticipated.

Ubicomp was made possible by the convergence of new advances in distributed and mobile computer systems, wireless communication devices, and new visualization technologies. With the disappearance of computers, we have witnessed the emergence of a new kind of ambient intelligence. Ambient intelligence refers to electronic environments that are sensitive and responsive to the presence of people (Friedewald et al., 2005). The aim is improving the quality of life by creating the desired environment and functionality via intelligent, personalized and interconnected systems and services. In an ideal ambient intelligent environment, the user is surrounded by a multitude of interconnected, invisibly embedded computer systems.

Ambient intelligence is capable of recognizing users, adapting to their preferences, and offering natural ways of interaction. The prototype project "Ambient Agoras" (*www.ambient-agoras.org*) already demonstrated that the computer as a device will disappear but its functionality will persist in a ubiquitous fashion. The Ambient Agoras environment was designed to transform places into social marketplaces of ideas and information ('agoras') and provide situated services, place-relevant information, and a feeling of the place ('genius loci'). It is possible to add new layers of information-based services to the place and 'memory' accessible to users. All these functions are achieved by integrating information into architecture via smart artifacts and expanding reality by providing better 'affordances' and information processing to existing places and objects. The emergence of ambient intelligence will represent a significant progress towards the development of information appliances that Norman (1998) envisioned.

Implications of Ubicomp for GIS

The development of GIS was made possible by digital technology, which has continued its growth according to its own logic as articulated in Moore's law, Metcalfe's law, and Reed's law. Undoubtedly, digital computing technology has influenced and will continue to influence the development of GIS. Up to now, the development of GIS was shaped largely by mainframes and personal computers – the first and second waves of computing. With the emergence of a third wave in computing, we must ask: How ubicom will affect/interact with GIS? Will ubicomp make GIS disappear and become part of the ambient intelligence? Are the GIS theories developed for the first and second computing paradigm suitable for a new geospatial future led by ubicomp?

Indisputably, ubicomp will embed more conventional GIS functions in various kinds of enhanced location-based services. Location-aware computing, or more broadly "context-aware" computing, will equip information infrastructure of all kinds with special capabilities to recognize and react to real-world contexts in which it operates. "Context" encompasses a plethora of factors, including user identity, time, physical location, and environmental conditions (such as local weather, etc.). The two most critical aspects of context are location and identity.

Context-aware computing systems respond to a user's location, either spontaneously (e.g. a friendly reminder of approaching a favorite restaurant, or warning of a nearby hazard) or when activated by the user's request (e.g. is this area prone to traffic jam or accidents?). With the continual growth of a variety of location-based services (LBS), many GIS functions that were previously confined to desktop PCs may become embedded as part of ubicomp or ambient intelligence.

The development of ubicomp will also entail another round of explosive growth of data of many kinds with greater spatial and temporal resolution. Real-time data acquisition (RTDA) will become a reality in the age of ubicomp, which will make just-in-time mapping (JITM) a routine operation in the near future. For example, the development of smart dust sensors – devices that combine microelectromechanical sensors with wireless communication, processing, and batteries into a package about a cubic millimeter in size – could potentially revolutionize RTDA and JITM.

Although currently still in their infancy, prototype smart sensors have demonstrated their capabilities to monitor environmental conditions that are hazardous to humans, determine the velocity and direction of passing vehicles in remote areas, or to be attached to humans, animals or even insects and record where they travel (Pister, 2002). The data transmitted wirelessly in real time from these sensors increase not only the volume but also the complexity of available geospatial data.

To store and analyze these colossal data sets obviously demands more storage space and computers more powerful than ever for data mining and research. So paradoxically, the age of ubicomp will also simultaneously create new demands for more powerful mainframe and personal computing. GIS theoretical foundations and functional capabilities in data acquisition, storage, analysis/modeling, and visualization need to be expanded to accommodate the new demands created by ubicomp.

Moving forward: Invisible GIS in/for the Post-human Age?

Along with bio- and nano- technology, geospatial technology has been regarded by the scientific community as one of the three defining technologies in the 21st century (Gewin, 2004). Although each of these three technologies will have far-reaching impacts on science and society, what will be really revolutionary is the convergence of two or more technologies, such as the development of locationally-responsive neural implants (by combing nano-, bio-, geo- and IT technologies,) context-aware computing (by integrating geospatial and IT technologies,), or smart dust as one component in a spatially-integrated monitoring system (by linking nano-, IT, and geospatial technologies.).

Ubicomp is a result of cross-fertilization among several different technologies. The development of GIS in the next few years will be riding on the crest of these converging technologies. To move GIS research forward, we must situate ourselves in the context of converging technologies and the expanding trends of interdisciplinary research (The U.S. National Research Council, 2003).

Back in 1997, Helen Couclelis wrote an article entitled "GIS Without Computers: Building Geographic Information Science from the Ground Up." Couclelis baffled many readers at that time because computing was on its way towards the peak of the second wave. To many of her readers, GIS without computers was still unthinkable.

Now, it is refreshing to read Couclelis' article in the context of ubicomp.

Although there are still some daunting technical challenges, computers are rapidly disappearing into the background/environment in the age of ubicomp. Indeed computers as the tool we have been using to study the world have become an integral part of the world itself. I believe a new GIScience research agenda "without computers" is precisely what is urgently needed. Such a research agenda for GIScience in the age of ubiquitous/pervasive computing would entail a major shift of our focus from predominantly technical issues to more substantive issues related to people, society, and environment.

Possible Research Questions for Invisible GIS in the Age of Ubicomp

Computational Questions:

How can one build up a ubiquitous computing infrastructure if there is none in the first place?

How do we transfer data seamless between different positioning systems?

What is the role of personal mobile devices in relationship to static and/or embedded public devices?

How do people migrate from explicit interfaces/interactions to implicit interfaces/interaction?

Spatial Questions:

Do we need new spatial models to represent data generated by ubicomp in real time and place?

What are the roles of distance in the age of ubicomp?

How will ubicomp affect business locational strategies?

How will ubicomp affect human settlement patterns and urban development from local to global scales?

Experiential/Behavioral Questions:

What are the basic models of interaction that promote coherence in individual as well as social experiences?

Will ubiquitous smart artifacts produce a more intelligent behavior?

How can people who inhabit these environments continually make sense of the world within which they live and what are the main technical and conceptual models needed to promote this?

How can people interact with invisible/disappearing computers/devices?

Will ubicomp promote more isolation/individualism or new forms of social interaction transcending age, gender, ethnicity, and class?

Social Questions:

What are the implications of putting sensors everywhere allowing to monitor, capture, record all kinds of interactions between people and machines/devices?

How can we ensure that privacy in future environments remains compatible with ambient intelligence?

Will ubicom become an always-on panopticon or a cooperation amplifier?

How to prevent tele-torture and geo-slavery in the age of ubicomp?

What new laws should be enacted to prevent the abuse of ubicomp?

Will ubicomp enlarge the digital divide or create digital dividends for all?

Environmental Questions:

What are environmental implications of ubicomp?

Will ubicomp reduce or enhance material, energy, and population flows?

Are we closer or further away from the goals of sustainability in the age of ubicomp?

How can ubicomp be deployed to advance the cause for the environment?

Will ubicomp sensors for animals, plants, and physical elements result in new types of environmental rights?

Clearly, there is no single discipline that could claim better or exclusive capabilities to tackle research questions listed in the table. To address these complex issues involving people, society and the environment would require a holistic perspective and close interdisciplinary collaborations. Harvey Miller's (2003) vision of a people-based GIS is certainly a significant step towards the right direction, but the difficulty with Miller's proposal is that the boundary between people and machines is no longer clear-cut (Adams, 2005). As Clark (2003) so forcefully argued, we are now all naturally born cyborgs, as bio-technological arrangements are emerging and evolving with a speed never seen before in history.

The merger of ubicomp with neural implant chips will make future computing not simply wearable, but implanted beneath our skin. Indeed, "the human skin is an artificial boundary: the world wanders into it, and the self wanders out of it, the traffic is two-way and constant" (Wolfe, 1952). The virtual, personal, social, and physical worlds are colliding, merging, coordinating, and reorganizing in many unpredictable and uncontrollable ways. If anyone wants a lesson in humility, he/she can try to make a prediction on what happens next.

Physicist Richard Feynman once said "I think I can safely say that nobody understands quantum mechanics (*www.physics.sfsu.edu/~jgregory/project_4/main.htm*). Considering not only what the converging technologies have done *for* us but also what they do *to* us, I'd like to venture that nobody knows for sure who we are and what we will become in the future as individuals, species, or society as a whole.

For lack of a better vocabulary, maybe the characterization of the emergence of a post-human age (Hayles, 1999) is not necessarily a bad idea. To me, during these early days of the ubicomp age, thinking about GIS without computers is not enough, and people-based GIS has already run into some conceptual difficulties. I concur with Rheingold (2002) that a new technological advance can be translated into a major social tsunami down the road.

Ubicomp, especially its further convergence with bio- and nano-technologies, will alter the human-nature, human-human, and human-self relationships in many fundamental ways. Is ubicomp going to become the so-called calm technology to reduce our anxiety as Weiser and Brown (1997) predicted, or will it become an intrusive technology that results in massive information anxiety and global environmental destruction as David Brin (1990) envisioned? Perhaps, more urgently than ever, we need to think about what we would like the invisible GIS of the post-human age to be.

Acknowledgement. I would like to thank (without implicating) Paul Adams, Jose Gavinha, and Bin Jiang for their perceptive comments on an earlier version.

(This column originally appeared in the January 2005 issue of Computers, Environment, and Urban Systems. *It is reprinted here with permission.)*

GIS As Media?

By Daniel Sui and Mike Goodchild

The dazzling development of GIS technology in recent years has rendered each of the traditional, mostly instrumental, views of GIS—as spatial database, mapping tool, and spatial analytical tool—inadequate to capture the fundamental essence of this technology and its social implications. Each year brings new software packages from innovative developers that are easier to use, more powerful, and more easily adopted by users with minimal training.

GIS and mapping tools are increasingly available on the Web, and an increasing number of sites offer advanced GIS services such as route finding and geocoding. In-vehicle navigation systems using GIS technology are becoming part of our daily lives (Cowen, 1994). In the next two years cell phones in the U.S. will be required to be geographically enabled—to be able to report their current location to an accuracy of 100m—in the interests of accurate response to emergency calls. New imagery is becoming available from commercial sources with spatial resolutions as fine as 1m, and is being distributed through new mechanisms such as distributed geolibraries and spatial data clearinghouses (NRC, 1993, 1999). New methods of data documentation are being used to support widespread sharing of spatial data via the Internet.

These new trends contrast sharply with the earlier view of GIS that prevailed into the early 1990s, as tools contained within a standalone computing system, serving the needs of their professional users by performing various forms of analysis that were too tedious, time-consuming, or expensive to perform by hand, on data collected and assembled for the purpose. These latest new developments in GIS have convinced us of the need of new conceptualization (or metaphors, for lack of a better term) for what GIS actually is and will become in the near future. We believe that the complex relationship between GIS and society can be better understood if you conceive of GIS as new media.

Media are generally understood as means of sending messages or communicating geographic information to the general public, and mass media are the instruments by which mass communication takes place in modern societies. Mass media are also the most effective means of broadcasting information to large numbers of people in a short period of time. In a very general sense, GIS can be understood as a new technological species in an already crowded media jungle, specifically focused on the communication of geographic information. The fact that GIS communicates geographic information in digital form merely illustrates its consistency with contemporary media, which now make widespread use of digital encoding at various stages.

Specifically, the meaning of "GIS as new media" can be broadly understood at least in the following dual senses. First, GIS have increasingly become a means to communicate certain aspects of the real world to the general public. Environmental Systems Research Institute (ESRI), Redlands, Calif., advertises its GIS software products as a "common language" for speaking about and "discovering" the world, while Intergraph, Huntsville, Ala., goes so far as to name its products "GeoMedia".

The role of communication is superseding the three traditional conceptualizations of GIS noted above (as tools for database management, mapping, and spatial analysis) because the impacts of GIS are realized only when results are presented to people on screens or on paper. Thus communication in its broadest sense must be put at the center when we discuss relationships between GIS and society—software vendors may have recognized this role of GIS better than many customers. Although data inventory and database management, automated mapping, and spatial analysis are still predominant tasks of most GIS projects, the goal of all GIS operations in the end is to communicate information to an audience in society. All GIS-related activities serve this final purpose—to communicate; the rest are intermediate steps serving primarily as means to communicative ends.

Second, significant changes have occurred in the past decade in the part played by digital geographic information in people's lives. Sites such as Mapquest (*www.mapquest.com*) provide millions of users every day with customized maps of routes and destinations, and sites such as Terraserver (*www.terraserver.com*) offer digital remote sensing imagery for sale to the general public. The Weather Channel and local television stations present forecasts that incorporate geographic information from satellite images, and display maps of forecasts from advanced computational models (Monmonier, 1996). Airline passengers are presented with dynamic maps of the flight's progress.

Humans have always needed to communicate geographic information, in describing discoveries, giving directions, or registering ownership. But recent technological developments appear to have opened a range of new possibilities. Indeed, with the emergence of digital places and digital individuals (Curry, 1998), we are witnessing not only Arnold Toynbee's etherization of history but also the etherization of geography.

It should be pointed out that the emphasis on the communication aspect of GIS has its parallels in cartography. Robinson and Petchenik (1975) first introduced Shannon's communication theory to understand the mapmaking process. Bertin (1978) made further elaborations on the linkages between the theory of communication and the theory of graphics. Martin (1996) tried to apply the Robinson-Petchenik communication model of cartography to understand GIS. By reconceptualizing cartography as a language for communication, Tobler (1979) introduced Chomsky's (1957) linguistic theory to develop a "transformation theory" of the cartographic process. Nyerges (1980) further explored Chomsky's "deep structure" vs. "surface structure" in both the cartographic and GIS contexts. Indeed, the communication paradigm even dominated the cartographic research literature in the 1970s and the 1980s (MacEachren, 1995). The recent literature on the basic issues of geographic information science (GIScience) also invoked various linguistic metaphors to study geographic information processing from the perspectives of communication and linguistics (Frank and Mark, 1991; Goodchild, 2000).

However all these previous works have engaged us in discussing technically oriented syntax issues (rules and grammars governing the relationships among signifiers); few have touched the complex and problematic semantic issues in GIS (the relationship between the signifier and the signified). Using communication theorist James Carey's words (Carey, 1989), all these previous works are dominated by the "transmission view" of communication at the technical level (how information is transmitted across space). What is lacking in the current literature is the "ritual view" of communication (how information is preserved in time) at the cultural and philosophical levels.

Reconceptualizing GIS as media opens new avenues for future research. As the impacts of media are much better understood today as a result of years of interdisciplinary research, GIS researchers can gain new insights on the social implications of GIS through the media perspective. For example, the tetradic analysis of GIS as media we conducted using McLuhan's framework has revealed some interesting hidden messages of GIS which defy their frequent dichotomous characterization as either good or bad (Sui and Goodchild, 2001). Instead, GIS transmit a very complex set of messages of simultaneous enhancement, obsolescence, retrieval, and reversal in society.

By reconceiving GIS as media, we can transcend the instrumental rationality currently rampant among both GIS developers and GIS practitioners and cultivate a more holistic approach to the non-linear relationships between GIS and society. The clear and present danger is not GIS or information technologies but our blissful ignorance of the implications of what they are going to do *to* us, because we have concentrated too much on what GIS can do *for* us. The only sensible way of dealing with this inherently technological ambivalence is to treat GIS media always as means to higher social ends.

Acknowledgment: Helen Couclelis, Jonathan Phillips, Paul Adams, and Daniel Overton provided critical comments on an earlier version of this paper. The Varenius project and the Alexandria Digital Library are supported through cooperative agreements between the University of California, Santa Barbara, and the U.S. National Science Foundation.

(This column originally appeared in the July 2001 issue of the International Journal of Geographical Information Science. *It is reprinted here with permission.)*

Section 6.
References

Adams, P.C. 2005. *The Boundless Self: Communication in Physical and Virtual Spaces*. Syracuse University Press, Syracuse, New York.

Bertin, J. 1978. "Theory of Communication and Theory of the Graphic." *International Yearbook of Cartography*, Vol. 28, pp. 118-126.

Brin, D. 1990. *Earth*. Bantam Books, New York.

Carey, J.W. 1989. *Communication as Culture: Essays on Media and Society* Unwin Hyman, Boston.

Chomsky, N. 1957. *Syntactic Structures* Mouton, The Hague.

Clark, A. 2003. *Natural-born Cyborgs: Minds, Technologies, and the Future of Human Intelligence*. Oxford University Press, New York.

Constanza, R., et al. 1997. "The Value of the World's Ecosystem Services and Natural Capital," *Nature*, Vol. 387, pp. 253-260.

Couclelis, H. 1997. "GIS Without Computers: Building Geographic Information Science from the Ground Up," in Kemp, Z. (ed.), *Innovations In GIS - 4: Selected Papers from the Fourth National Conference on GIS Research UK (GISRUK)*, pp. 219-26. Taylor & Francis, London.

Cowen, D.J. 1994. "The Importance of GIS for the Average Person." *Proceedings of the First Federal Geographic Technology Conference,* pp. 7-11, GeoTec Media, Boulder, Colo.

Curry, M.R. 1998. *Digital Places: Living with Geographic Information Technologies* Blackwell, London.

Daily, G. (ed.) 1997. *Nature's Services: Societal Dependence on Natural Ecosystems*. Island Press, Washington, D.C.

Dobson, J.E. 1990. "The Two Faces of Geography," *Mensa Bulletin,* Vol. 337, No. 1, pp. 3-4.

Dobson, J.E. 1994. "GPS Advances Geography's Field Tradition," *GIS World,* Vol. 7, No. 1, pp. 60-61.

Dobson, J.E. 1994. "The GPS Field station Rides Again," *GIS World,* Vol. 7, No. 9, p. 54.

Dobson, J.E. 1996. "Real-Time GIS Is on the Rise," *GIS World,* Vol. 9, No. 9, p. 34.

Dobson, J.E., Bright, E.A., Coleman, P.R., Durfee, R.C., Worley, B.A. 2000. "LandScan: A Global Population Database for Estimating Populations at Risk," *Photogrammetric Engineering & Remote Sensing,* Vol. 66, No. 7, pp. 849-857.

Dobson, J.E., Bright, E.A., Ferguson, R.L., Field, D.W., Wood, L.L., Haddad, K.D., Iredale, III, H., Klemas, V.V., Orth, R.J. Thomas, J.P. 1995. *NOAA Coastal Change Analysis Program (C-CAP): Guidance for Regional Implementation. National Oceanic and Atmospheric Administration Technical Report NMFS 123.* U.S. Department of Commerce, National Technical Information Service, Springfield, Va.

Dobson, J.E., Rush, R.M., Peplies, R.W. 1990. "Forest Blowdown and Lake Acidification," *Annals of the Association of American Geographers,* Vol. 80, No. 3, pp. 343-361.

Frank, A.U., Mark, D.M. 1991. "Language Issues for GIS," pp. 147-163 in *Geographical Information Systems: Principles and Applications*, (eds.) Maguire, D.J., Goodchild, M.F. Rhind, D.W. Longman Scientific and Technical, Harlow, United Kingdom.

Friedewald, M., Da Costa, O., Punie, Y., Alahuhta, P. Heinonen, S. 2005. "Perspectives of Ambient Intelligence in the Home Environment." *Telematics and Informatics,* Vol. 22, pp. 221-238.

Gates, B. 2003. "The Disappearing Computer." Available online, *www.microsoft.com/presspass/ofnote/11-02worldin2003.asp*.

Gewin, V. 2004. "Mapping Opportunities." *Nature*, Vol. 427, pp. 376-377.

Giovinetto, M.B., et al. 1990. "Dependence of Antarctic Surface Mass Balance on Temperature, Elevation, and Distance to Open Ocean," *Journal of Geophysical Research & Atmospheres*, Vol. 95, pp. 517-531.

Goodchild, M.F. 2000. "Communicating Geographic Information in a Digital Age." *Annals of the Association of American Geographers*, Vol. 90, No. 2, pp. 344-355.

Hayles, N.K. 1999. *How We Became Posthuman: Virtual Bodies in Cybernetics, Literature, and Informatics*. University of Chicago Press, Chicago.

Hightower, J., Borriello, G. 2001. "Location Systems for Ubiquitous Computing." *IEEE Computer*, Vol. 33, No. 8, pp. 57-66.

MacEachren, A.M. 1995. *How Maps Work: Representation, Visualization, and Design*. Guilford Press, New York.

Mann, S. 1996. "Smart Clothing." Available online at: *http://wearcam.org/smart_clothing*.

Martin, D. 1996. *Geographic Information Systems: Socioeconomic Applications*. (second edition) Routledge, London.

Miller, H.J. 2003. “What About People in Geographic Information Science?” *Computers, Environment and Urban Systems*, Vol. 27, pp. 447-453.

Monmonier, M. 1996. “Electronic Weather Maps,” in Hanson, S. (ed.) *Ten Geographic Ideas that Changed the World*, pp.56-69. Rutgers University Press, New Brunswick, NJ.

National Research Council (NRC). 1993. *Toward a Coordinated Spatial Data Infrastructure for the Nation* National Academy Press, Washington, DC.

National Research Council (NRC). 1999. *Distributed Geolibraries: Spatial Information Resources* National Academy Press, Washington, DC.

Norman, D.A. 1998. *The Invisible Computer: Why Good Products Can Fail, the Personal Computer is so Complex, and Information Appliances are the Solution*. MIT Press, Cambridge, Mass.

Nyerges, T.L. 1980. *Modeling the Structure of Cartographic Information for Query Processing*. Unpublished Ph.D. dissertation. Ohio State University, Columbus, Ohio.

Pearce, D.W. 1993. *Economic Values and the Natural World*. MIT Press, Cambridge, Mass.

Pimm, S.L. 1997. “The Value of Everything,” *Nature*, Vol. 387, pp. 231-232.

Pister, K. 2002. “Smart Dust: Autonomous Sensing and Communication in a Cubic Millimeter.” Available online at *http://robotics.eecs.berkeley.edu/~pister/SmartDust*.

Platt, R.S. 1959. “Field Study in American Geography: The Development of Theory and Method Exemplified by Selections.” University of Chicago, Department of Geography, Research Paper No. 61, Chicago.

Rheingold, H. 2002. *Smart Mobs: The Next Social Revolution*. Perseus Publishing, Cambridge, Mass.

Robinson, A.H., Petchenick, B.B. 1975. “The Map as a Communication System.” *The Cartographic Journal*, Vol. 12, pp. 7-14.

Satyanarayanan, M. 2001. “Pervasive Computing: Vision and Challenges.” *IEEE Personal Communications*, Vol. 8, No. 4, pp. 10-17.

Sui, D.Z., Goodchild, M.F. 2001. “A Tetradic Analysis of GIS and Society Using McLuhn’s Law of the Media.” *Canadian Geographer*.

The U.S. National Research Council. 2001. *Embedded, Everywhere: A Research Agenda for Networked Systems of Embedded Computers*. National Academy Press, Washington, DC.

The U.S. National Research Council. 2003. *IT Roadmap to a Geospatial Future*. National Academy Press, Washington, DC.

Tobler, W.R. 1979. "A Transformational View of Cartography." *The American Cartographer*, Vol. 6, pp. 101-106.

Weekly World News. 1993. "Half-Alligator, Half-Human Found in Florida Swamp!" *Weekly World News*, November 9, 1993, p. 1.

Weiser, M. 1991. "The Computer for the Twenty-first Century." *Scientific American*, Vol. 265, pp. 94-104.

Weiser, M. 1993. "Some Computer Science Issues in Ubiquitous Computing." *Communications of the ACM*, Vol. 36, No. 7, pp. 75-85.

Weiser, M., Brown, J.S. 1997. "The Coming Age of Calm Technology," in Denning, P.J., Metcalfe, R.M. (eds.), *Beyond Calculation: The Next Fifty Years of Computing*. Copernicus, New York.

White, R.W. 1994. "A Few Good Gatormen: Beating the Swamps for Mythological Survivors." *Outside*, March, pp. 39-43.

Wolfe, B. 1952. *Limbo*. Ace Books, New York.

Section 7. Digital Earth and the Impact of GIS

Introduction

By Matt Ball

Former Vice President Al Gore laid out his vision for a "Digital Earth" in January 1998. The vision discussed the ability to "virtually fly through space down through progressively higher-resolution datasets to hover above any point on Earth's surface—and then display information relevant to that location from an infinite number of sources."

Dobson's paper, which was presented at the International Symposium on Digital Earth (1999), embraces this vision and assesses the amount of work needed to overcome technical obstacles. Dobson's description of a space/time/entity/process (STEP) system to analyze spatial phenomena across disparate fields points to a new level of interdisciplinary integration that closely models the role of the geographer.

Waters' column, "Does GIS Make Us Smart?" explores the interface issue between the tool and its purpose. He states, "too much time is spent on learning the mechanics of the GIS and too little on reflecting about what we are doing and why we are doing it." In his next column in the series, Waters sees promise in 3-D visualization as a means of "virtual World delivery via the Web." This creates a more-organic interaction than a purely desktop product.

The combination of Google's geographic search capabilities, and its geographic exploration system platform Google Earth, is the focus of Sui's column, "What Can We Learn from Google's Success?" Sui notes that "considering the fact that each entity in the virtual and physical world can and will be georeferenced and tracked with explicit geographic information, the entire world will eventually become a giant database to be searched."

Waters notes the importance of a temporal component to this search capability to track changes through time. As it stands now, the move to digital data means a loss of the archival capacity of mapping to mark change over time.

The advent of geographic exploration systems broadens the scope of geospatial technology use by providing a simpler interface for geospatial data exploration as well as a compelling 3-D environment. They build on decades of geographic information system (GIS) development by taking advantage of the geospatial data created by GIS practitioners.

This book ends with Dobson's whimsical piece, "Geographer Questioned at Pearly Gates." This ultimate accounting of a life's work describes several misconceptions about the work that geographers and GIS practitioners undertake. Certainly some of these misconceptions will resonate with you, and perhaps this dialog will help pave your way to a brighter path.

Geographic Exploration of Digital Earth

By Jerry Dobson

I commend the conveners of the 1999 International Symposium on Digital Earth, United States Vice President Albert Gore, Mike Goodchild, Tim Foresman, and others around the world who have created and championed the concept of Digital Earth. The name itself evokes the image of a complete representation of the earth in a more realistic, comprehensive, integrated, and manipulable form than traditional globes and global models. It is almost as if we could take this wonderful world and hold it in our hands, watch it spin, and scrutinize any part of it to our hearts content.

Indeed, when implemented, Digital Earth will allow analysts and decision makers to view a spinning globe; select and zoom to any place (eg., continent, region, country, city, or point); and instantaneously access crucial information about that place. For those who prefer visualization or virtual reality, it will be possible to fly-through or walk-through any landscape on earth at any reasonable scale, and the definition of what is reasonable grows finer every year. For those who need to conduct geographic and other scientific analyses, it will be possible to model the atmosphere, oceans, terrestrial surface, and inner Earth at any reasonable scale. We will be able view and study the earth realistically with satellites orbiting, atmospheric conditions changing, and surfaces selected for the purpose at hand. The time scale can vary, too, as clouds form, nations evolve, or continents drift.

The technical challenges will be enormous. Fundamental technology development will include improved graphics, visualization, image processing, spatial data structures, computer cartography, and support of global and regional modeling. Geo-visualization and 4-D simulation will require parallel processing, high-performance (tera flop) computing architectures and advanced (100-plus terabyte) storage systems. Even the most powerful computers and storage devices will require optimization of software functions and maximal data compression. A new class of space/time/entity/process (STEP) linkage systems must be invented for representing and analyzing spatial phenomena across fields as diverse as medicine, biology, engineering, and physics. The resulting system will be a substantial improvement over current global models and will set the standard for the next generation of geographic technology. It will differ from current technology in that it will be supported with an extensive, high-resolution data collection; will overcome significant barriers to the integration of disparate types of geographic data (including both raster and vector data); and will be designed to facilitate modeling and other types of analyses as well as visualization, virtual reality, and cartographic display.

Digital Earth will be especially useful in support of scientific research on major global issues such as (a) the search for causal mechanisms in continental drift and plate tectonics and (b) the attempt to understand climate change and its impacts on human systems (eg. redistribution of agricultural production, income, and energy demand). Yet, it will be equally well suited for detailed analyses of land cover change (eg. forests, wetlands, deserts); sources and effects of atmospheric pollution; energy assessment and planning; emergency management; commercial and military logistics; epidemiology; regional economics; and resource management.

The Age Old Quest to Explore The Earth's Frontiers

Now that our spirits have soared with unbounded possibilities, let's settle back to plain old analog earth in its present state of affairs, and ask what's missing. Other speakers will address the satellites and sensors that are needed to acquire imagery at finer spatial resolution, greater spectral definition, and improved temporal frequency. Others will address computational and analytical needs. Perhaps someone will address information that cannot be acquired through remote sensing and thus requires field investigation or other traditional data collection. I, too, am concerned about these matters, and my colleagues at Oak Ridge National Laboratory (ORNL) are working diligently to solve those for which we have funding and expertise. But there is a greater societal issue that speaks to human aspirations in the coming millennium and to which I address my remarks today: Is geographic exploration still necessary in the age of Digital Earth?

The rhetoric of geographic information science, including Digital Earth, often implies that future knowledge will be generated, not from exploration of the unknown but from better representation, integration, and analysis of what is already known. More broadly, there is a tacit societal presumption that the Age of Exploration has ended once and for all, that there is, in essence, nothing left to discover. Science itself has become a plodding quest for incremental knowledge in which quantum discoveries are unexpected and often unwelcome. No wonder the public has lost interest in what we learn, and support for scientific research is on the decline.

A recent review in *The New Yorker* (Hertzberg 1999) illustrates society's typical attitude toward geographic exploration. Hendrik Hertzberg belittled the National Geographic Society's *Adventure* magazine, not for its writing or photography or style, but rather because, the exploration of unknown territory, conducted of necessity under conditions of extreme hardship is obsolete. That sounds like what Columbus contemporaries were saying in 1491, and it's simply not true. Yes, the nature of exploration has changed, but there is still much left to discover. In the earlier Age of Exploration it was a stunning accomplishment sim ply to observe a new landmass and report its existence. Today, exploration is more demanding than ever as we probe places that are deeper (oceans floors, caves, inner Earth) and higher (atmosphere, planetary bodies), search for tangible evidence of historic and pre-historic landscapes, and seek new geographic understanding of complex earth processes. Through technology, exploration can now be extended to places where humans cannot go, to phenomena that cannot be observed directly by human senses, and to macroscopic features so large they can be observed and investigated only through remote sensing, GIS, and advanced geo-visualization. In this new age, Digital Earth is an essential tool for discoveries that might not have been recognized before, even if the observations had been collected.

Recently, the council of the American Geographical Society (AGS) voted to revitalize the Society's historic mission to support geographic exploration, and I was named Director of Exploration. In this new capacity, I have come face to face with society's initial skepticism. The bad news is that skepticism is pervasive. The good news is that most people quickly reconsider their views when the new nature of exploration is explained and a few examples of recent discoveries are offered.

20th Century Exploration

We at the AGS have in our possession a wondrous globe that extols the history of geographic exploration in the Twentieth Century. It is called The Flier's and Explorer's Globe or sometimes simply The Flier's Globe. It has belonged to the AGS since 1929. Its extraordinary value lies in what's been scribbled and scrawled across its surface.

Picture, if you will, an antique globe with Charles Lindbergh's route etched across the Atlantic and signed in his own hand. How much would such a collectible bring at Sotheby's? How much more if Amelia Earhart also signed and marked her route as the first woman to fly across the Atlantic in 1932? Now, add Wiley Post, first man to fly solo around the world (1933), and Harold Gatty, who helped him set an around-the-world speed record shortly before Post and Will Rogers died on the same route. Add Louise Boyd, the most accomplished of early women aviators. Add John Glenn, first person to orbit the earth (1962), with a photograph of President John Kennedy and Vice-President Lyndon Johnson watching him sign. Add astronauts Frank Borman and James Lovell.

That's why it's called The Flier's Globe, but what about explorers? Look north, and you'll find Robert Peary, president of the Society (1903 to 1907) and first person to reach the North Pole (1909). Look south, and you'll find Roald Amundsen, first to reach the South Pole (1911). Elsewhere, you'll find Fridtjof Nansen, who led a lengthy Arctic expedition (1893-1896), directed the League of Nations refugee relief effort after World War I, and won the Nobel Peace Prize in 1922; Richard Byrd who explored the Arctic and Antarctic regions and established the Little America research base in Antarctica; and Sir Hubert Wilkins, among the first to explore the Arctic by air. You'll even find Sir Edmund Hillary's signature on Mt. Everest and William Beebe's signature near Bermuda where he and Otis Barton set a depth record of 3,028 feet in the Bathysphere (1934).

The Fliers Globe contains these signatures and many others for a total of 62 individuals, and some names are repeated for more than one feat. The Fliers Globe is not a complete record of 20th Century exploration, however, for we somehow missed a few obvious candidates. Jacques Piccard and Don Walsh, for example, descended to the bottom of the ocean (35,800 ft.) in the Bathyscaphe in 1960, but their names do not appear over the Marianas Trench. Jacques Cousteau's name doesn't appear although he certainly qualified on the basis of many ocean explorations. Astronaut Neil Armstrong definitely should have signed when he returned from the moon in 1969. Cosmonaut Valentina Vladmirovna Tereshkova should have signed as the first woman in space in 1963. Some, like Cousteau, are lost forever, but the Society has begun a process to remedy as many oversights as possible.

Are we, in the age of Digital Earth, still accomplishing feats of comparable magnitude? I've asked friends and colleagues which modern explorers and aviators are worthy to sign. Robert Ballard has been suggested, as have Sylvia Earle and Graham Hawkes for their separate and joint accomplishments in deep ocean exploration. Dave Allured, Neil Backstrom, and Rick Bridges clearly qualify for discovering the world's deepest cave, Lechuguilla in New Mexico (1986). For aviation, Bertrand Piccard (son of Jacques Piccard) and Brian Jones have been suggested for their unprecedented circumnavigation of the earth by balloon. It's comforting to know the quest continues in both arenas.

Two other candidates definitely would qualify as explorers, if only they were human. NASA's Mars Rover and Hubble Telescope have explored distant places in the old fashioned sense of being there and recording what they saw, albeit digitally and remotely. Herein lies a hint of things to come as being there becomes less important (and less effective) than probing and reporting what lies beyond our human grasp.

Remote Exploration

The roots of remote exploration go back farther than one might think. As early as the 1860s, there was a serious scientific proposal to explore Africa by balloon. The geographically gifted young visionary failed to attract funding, however, and later yielded to a publisher's suggestion to rewrite the concept as science fiction. Thus, *Five Weeks in A Balloon* (Verne 1863), a fictional account of scientists using a balloon to explore Africa, became Jules Verne's first novel. Today, Verne (1828-1905) is famous for predicting airplanes, submarines, television, guided missiles, spacecraft, and satellites almost a century before their actual invention. Viewed as a geographer, he was well-grounded in the discipline and his subjects covered the same frontiers that challenge us today—space, oceans, inner Earth.

In 1904, Professor William Morris Davis, a key founder of modern geography, wrote to the AGS urging the Society to fund two exploration projects. One was a traditional expedition to learn more about glaciation of the Andes Mountains. The other was a steamer voyage to collect kite soundings of the atmosphere over the tropical Atlantic Ocean. The latter is clearly analogous to the types of remote exploration that are conducted today in space, under the sea, or deep into the inner Earth. The move toward remote exploration thus began even before the last terrestrial frontiers were breached at the North Pole (1909) and South Pole (1911). Indeed, the movement actually began hundreds of years earlier if ocean depth soundings are considered in the same light.

Aquaterra, The Next Frontier

In January 1999, I called for the exploration of Aquaterra the distinctive coastal and littoral zone that was inundated and exposed repeatedly during the Pleistocene ice ages and now rims all the world's oceans to a depth of about 125 meters and to a height of about 5 meters above current sea level (Dobson 1999). In August 1999, the *Professional Geographer* (1999) published a special issue on the geography of ocean space. In that issue and elsewhere for several years, Dawn Wright has championed the use of geographic information systems (GIS) to represent and analyze oceans and associated geophysical phenomena. Clearly, there is interest among geographers in exploring the oceans just as we explored land areas during the Age of Exploration. I am firmly convinced that first Aquaterra and then the Deep Ocean must be viewed as the next final frontiers of Digital Earth.

Aquaterras defined space is a function of time, bounded by the highest and lowest sea levels of the past 120,000 years. This covers the four best-documented cycles of rise and fall and also the entire span of time during which anatomically modern humans are known to have existed. Excellent data are available for the most recent sea-level rise, based primarily on analysis of Barbados corals which indicate lows of -120 m at 17,000 years (Fairbanks 1989) and -125 m at 21,800 years (Peltier 1994). Earlier fluctuations are inferred from heavy oxygen isotope (18O) concentrations in marine sediment cores (Stringer and Gamble 1993) and from the advance and retreat of glaciers.

A long-term goal and major challenge of exploration must be to regionalize Aquaterra with reference to three-dimensional space and time. At any given moment, Aquaterra is either fully inundated during periods of maximum transgression as it nearly is today; fully exposed as it was during the maximum regression 21,800 years ago; or partially inundated and partially exposed. Regionalization, thus, depends on creating a reliable chart of dates and corresponding mean sea levels during the past 120,000 years. It is essential, however, to advance far beyond mere outline. Next, we must map and describe its physical geography, emphasizing the lands exposed at any given time. Where were Aquaterra's rivers and estuaries? Surely, they were focuses of biological activity and human settlement then as in later historic times. Where were its up lands? Surely, elevated areas would have served humans and animals alike as refuges from flooding, and the theoretical maximum local relief *within* Aquaterra is 130 m. Where were its distinctive landforms? Waterfalls twice as high as Niagara conceivably could have existed there and could have been valued in religious or secular ways, but no one has searched for them.

The exploration of Aquaterra will be a technological exploration unlike any before on earth, at least as challenging as space exploration of the moon and planets. Unlike the moon and Mars, there will be no direct viewing of the whole, because the field of view for each observation is so limited. Thus, Digital Earth will be the macroscope through which we view the whole, as well as the microscope through which we examine each piece. Geo-visualization will be paramount. No one in our lifetime will ever see Aquaterra, as a whole, except through the lens of Digital Earth.

A Call For Exploration

In 1903, Admiral Robert E. Peary, then president of the AGS, urged the Society to adopt an aggressive policy of polar exploration. We need a large and vigorous policy, he said. "Of late years exploration has of necessity been a work of large details, opening the hearts of continents and pushing northward and southward, till today only the northern and southern apices of the earth still hide in the mists and gloom of the polar nights. If we wish to keep in the lead and be in at the death in the final geographical conquest of the world, our first efforts must be in those two directions, north and south."

He urged the Council to purchase a ship and carry on original scientific investigations (explorations, surveying, hydrography, deep-sea dredging, etc.) (Wright 1952, p. 138). I urge the same for Aquaterra, a global feature so forgotten by mankind that Admiral Peary failed to mention it in his remarks. I address my call to the conveners of this symposium, the AGS, the National Aeronautical and Space Administrations (NASA) Digital Earth Program, the National Oceanic and Atmospheric Administration (NOAA), and anyone else who cares deeply about geographic exploration.

(This paper first appeared in Towards Digital Earth, *proceedings of the international symposium on digital earth, Science Press, 1999. It is reprinted here with permission.)*

Does GIS Make Us Smart?

By Nigel Waters

Norman's back! Donald Norman is the cognitive psychologist who wrote the book *The Design of Everyday Things* (1988). His main argument was that things that we use in our daily lives should be designed to reflect the manner in which we use them. This "people-centered" philosophy is reflected in his most recent book, *Things That Make Us Smart*.

Things That Make Norman Smart

Norman (formerly head of the cognitive science department at the University of California, San Diego, and now an Apple Fellow at the company's headquarters in Cupertino, Calif.) emphasizes that he is a technological optimist. He distinguishes himself from technological pessimists such as Jerry Mander who (1992) offered some of the harshest criticism to date of the computer revolution. By contrast, Norman believes that, while all is not well with the revolution, redemption is still possible if those who are responsible for the design of technology adopt a human-centered methodology. So far that only has occurred in a spasmodic fashion. More commonly technologists appear to adopt the motto of the 1933 Chicago World's Fair: "Science Finds, Industry Applies, Man Conforms."

The leitmotif of Norman's work is the distinction between reflective and experiential cognition is knowing what to do, automatically, without thinking about it. Experiential cognition is what Indy champion Nigel Mansell has when he moves his racing car through a chicane (Spackman and Tan, 1993). Reflective thought might occur in situations where Mansell considers whether to change to wet tires from slicks or whether to pit for one more fuel stop. In GIS it is the difference between running the system for some routine task or for concerning oneself with the design of new procedures.

The modes of thought often are mixed, and one might consider a continuum from extreme experiential modes such as playing a Nintendo game, through operating a GIS, to writing macros and finally to redesigning the conceptual structure of a GIS. The latter then represents the reflective end of the continuum.

Norman (1993) argues that the greatest peril in the design of today's technology is that of "experiencing when one should be reflecting" (p. 27), and that seems to be a particular concern with the design of GIS. Too much time is spent on learning the mechanics of the GIS and too little on reflecting about what we are doing and why we are doing it. Some universities, such as Sherbrooke (Michel Rheault, Dept.

of Geography, University of Sherbrooke), have gone so far as to separate (in different courses) the mechanics of learning to operate a GIS from the more intellectual aspects of understanding and reflecting upon the applied and conceptual issues.

Helping GIS Make Us Smart

The question then is how can we design GIS to make us smarter? How do Norman's ideas help? According to Norman (1993), our tools should be invisible—transparent to the user. Two difficulties may arise: the tools themselves may be designed inappropriately, or they may be used in inappropriate ways.

Some tools should be used experientially, without a second thought, but instead they require and demand reflection. The original car cell phones were a design disaster. They required you to remove your hands from the steering wheel—not when you felt it was safe (like a car radio), but whenever a call came in. Such phones were replaced by phones that were almost "hands-free." But the ideal would be a voice-activated phone, and that is the approach that should be used for GIS-based, in-vehicle navigation systems. In short the map metaphor used in such systems is an accident waiting to happen. Airline pilots have heads-up displays, but that technology would be too expensive for cars. The best solution would again be a voice-activated system (McCrone, 1993). You tell your computer your present intersection and your destination and it tells you where to go. It could even issue the turns and directions in a paced fashion that considers your speed. Wouldn't that make you a smart driver?

Other tools should be used in reflective mode, yet they do not allow us to make comparisons and explorations of alternatives, or to reflect on the quality or sensitivity of solutions. That is the primary goal of spatial decision support systems, which employ many of the approaches developed in operations research. The most recent trend in that discipline is "what-if scenarios." In the past, researchers solving location problems were content to determine the optimum or best solution, because that was exactly what bureaucratic and political decision makers wanted—one, single, perfect answer to the problem. Today such decision makers are interested in exploring suboptimal solutions that may be attractive for other, non-quantifiable reasons.

GISs are becoming Spatial Decision Support Systems (SDSSs), but they are not there yet. An ideal SDSS designed, say, for transportation purposes might allow the planner to explore the effects of different attribute interactions associated with the network. For example, suppose there is a left turn attribute associated with a given link of a high-speed, in-city freeway. The GIS/SDSS operator might want to simulate the probability of a following vehicle driving into the back of the turning vehicle. Even more effective would be a system that would flag automatically any such link that had a speed limit higher than 30 m.p.h. Wouldn't such innovations make you a smart transportation planner?

Norman's person-centered motto for the 21st century (in which the logic of the Chicago World's Fair motto is reversed) provides a far more lofty ideal for those concerned with GIS design: "People Propose, Science Studies, Technology (GIS) Conforms."

(This column originally appeared in the June 1994 issue of GIS World.)

Is GIS Prepared for the Virtual Reality Age?

By Nigel Waters

This column was inspired by Michael Schrage's keynote address at GIS/LIS '96 in November 1996 in Denver and by a colleague complaining about how hard it is to keep up with the changing technology in computer mapping.

Schrage's talk, "Multimedia, Hypermaps, and Cartopreneurs: Mapping the Future of the Marketplace," was great news for geographers. He told us that mapping and GIS on the Internet were going to be a multibillion-dollar business by 2000 [Update - AOL's purchase of MapQuest in December 1999 for $1.1 billion was a notable milestone in this online mapping growth, and a fulfillment of this prophesy].

He told us that geographers owned all of the Internet metaphors, which are entirely spatial. Consider Mountain View, Calif.-based Netscape Communications Corp.'s Netscape Navigator; Redmond, Wash.-based Microsoft Corp.'s Internet Explorer; and the whole idea of virtual worlds. Computer magazines provide Internet maps, and Schrage suggested that we'll soon be able to organize our Web bookmarks spatially as well.

He told us that newly minted Ph.D.s in GIS-based geography would command annual salaries of $100,000. He harped on the fortunes of geographers so much that at the talk's conclusion, the fellow in front of me muttered irritably that "GIS wasn't just about geography."

The utility of spatial metaphors has been acknowledged in academic articles (Skupin and Buttenfield 1996) and in popular, "high-tech" literature (Stephenson 1993). How do we create new virtual spaces, places, worlds and geographies on the Internet? The answer is Virtual Reality Modeling Language (VRML).

The VRML 2.0 Standard

VRML, a file format created by Silicon Graphics Inc. (SGI), Mountain View, Calif., for describing interactive 3-D worlds that can be viewed across the Internet, is now in its second revision. Development process details for the VRML standard and other relevant Internet resources may be found at *www.web3d.org*.

So what can you do with VRML? You can do the old stuff—create 3-D worlds. When I first started browsing through the *VRML 2.0 Sourcebook* (Nadeau et al. 1996), I had a sense of *déja-vu*. The authors discuss such topics as predefining shapes; 2-D and 3-D coordinate systems; the right-hand rule for 3-D; interpolation; elevation grids, now called Digital Elevation Models (DEMs); the use of color; and transformations such as scalings, rotations and translations.

I taught these topics in the late 1970s and early 1980s in my computer mapping course at the University of Calgary, Alberta, Canada. As that course evolved into a GIS course, I incorporated these topics, and they were included in the core curriculum of the National Center for Geographic Information and Analysis (NCGIA), University of California, Santa Barbara (Kemp 1996). My colleague who complained of the many new things to learn in computer mapping may well be able to use many of his old ideas.

But VRML 2.0 does much more. In addition to the aforementioned capabilities, it allows the developer to animate objects in the virtual world, to incorporate sounds and movies and to make the world interactive—it's a sensory-rich environment (Nadeau et al. 1996). We can use VRML 2.0 to build real and virtual geographies and deliver them across the Web. Web sites that describe how to download VRML browsers for various platforms are available from the *VRML 2.0 Sourcebook*.

VRML 2.0's Impact on GIS

The impact on GIS of such a sophisticated, interactive language that allows virtual world delivery via the Web is incalculable. For GIS education, VRML will allow simulation of processes in three dimensions, dynamically and with sound. New curricula will be designed based on this technology—the new NCGIA core curriculum will be available on the Internet (Kemp 1996). Virtual field trips have become popular in some geography courses. I've used Jeremy Crampton's widely praised "field trip" to Bosnia in an introductory university geography course. But this trip would be more interesting if we could fly through a DEM of Bosnian countryside or hear interviews with citizens and soldiers on either side of the conflict.

A critical question is how quickly VRML 2.0 will execute standard GIS operations. In our experience at the University of Calgary with interactivity using the Java programming language developed by Sun Microsystems, Inc., Mountain View, Calif., VRML 2.0 is disappointingly slow (Bertazzon and Waters 1996). With faster processors and more efficiently written programs, however, VRML may become an important part of GIS' future.

[Update – VRML 2.0 became the platform for the ISO standard VRML97, however each VRML vendor implemented a different subset of the spec, and it never gained traction. VRML lives on in its XML flavor, X3D, which has grown to encompass VRML's siblings H-Anim (Humanoid Animation) and GeoVRML. See *www.web3d.org* for details on the current standard. Details on the history and evolution of VRML can be viewed at *www.mitra.biz/vrml/vrml2/mw_history.html*.]

(This column originally appeared in the March 1997 issue of GIS World.*)*

Research Under Way on GIS, Memory Aides and Intimate Technology

By Nigel Waters

One of the more intriguing developments in computing to become popular in recent years is the memory aide. Technophiles now routinely consult their electronic organizers. Desktop computer aficionados are lost without their Personal Information Managers. [Update: In February 2006 we can recall the panic felt by the U.S. federal government when a patent infringement lawsuit threatened to cause a blackout of Waterloo, Ontario-based Research In Motion's (*www.rim.com*) Blackberry service.]

Why is that? In part, I believe it is due to the increasing complexity of our lives, information overload and the so-called information explosion (Wurman, 1989). Moreover, it may be due to the failing of short-term memory as we age. Whatever the reason, these aides are becoming increasingly important to us. As McCrone (1994) notes, we assign much of our identity to our memories.

That point makes for an amusing debate between Ebenezer Cooke and Henry Burlingame in John Barth's infamous novel *The Sot-Weed Factor* (1960, pp. 138-142). Cooke asserts that memory is the "House of Identity, the Soul's dwelling place." But Burlingame presents four obstacles to placing human memory on such a lofty pedestal. First, there are breaks, or lapses, in memory—we forget things. Second, if memories disagree, we have no way of ascertaining who is right without resorting to some outside help. Third, memories are selective. We remember what we want and may conveniently forget the rest. [Update: Research reported in the journal Scientific American Mind (December 2005) has suggested that Barth may have had it wrong and that it's the persistent bad memories that are the most lasting and difficult to erase—without a pill or two.] Fourth, memories are biased. We frequently don rose-colored glasses when it suits us. (Burlingame brings up a fifth concern that is more metaphysical, so let's leave it alone.) Research is under way to help resolve these human frailties.

Memory Aides Today

Perhaps the most innovative research on such topics is being carried out by researchers at Xerox's EuroParc research lab in Cambridge, England. A detailed discussion of various developments is provided by McCrone (1994). Several computerized systems have been under development for the last six years, including the following:

• No Time used an electronic notebook and camcorder system to keep a time-stamped record of meetings, allowing cryptic jottings to be "clicked-on" to reveal a detailed transcription of events. Thus, disagreements over the exact record of the meeting could be resolved quickly, thereby overcoming Burlingame's second and perhaps third objections to human memory.

• The Pepys system, named in deference to Samuel Pepys, the father of all diarists, can be used to generate electronic diaries. The system used a combination of hardware and software. Each staff member was issued a smart card that acts as an identification badge and broadcasts the person's location to infrared sensors located in each room. (Transit systems used similar technology for bus tracking.) Thus, each individual was monitored in "Star Trek" fashion throughout the work day (Want et al., 1992).

Pepys was extra clever, because it attempted to infer purpose from the individual's movements and groupings and prepares an electronic diary/report for the following morning. A visual version, Vepys, which used video cameras to take snapshots of each individual every 10 minutes, is under development. Such systems would help us overcome memory lapses and the aforementioned selective biases noted by Burlingame. Diaries of behavior have been an important research tool in industrial engineering and behavior modification since the early days of time and motion studies. Vepys was likely to make their collection and analysis that much easier. [Update: It is worth noting that this early work is complementary to research currently being undertaken by Yuzuru Tanaka at the University of Hokkaido, Sapporo, Japan (*www.meme.hokudai.ac.jp/index-e.html*), and Klaus Jantke at Ilmenau in Germany (*www.tu-ilmenau.de/site/ifmk/mma.html*) on the Digital Bibliotheque City that provides a three-dimensional virtual interface to a digital library. The cell phone becomes the memory aide since it can be used to access the virtual library. I am collaborating with this research at the University of Calgary.]

• Forget-Me-Not acquired contextual information concerning background events such as phone calls, computer file accesses, and news and weather that could be used to supplement the information stored in Pepys-type systems.

• The Marcel System used a video camera to keep track of paper moving across a person's desk.

• "Awareness Servers" were described by McCrone (1994) as just one example of the so-called "intimate" technology now under development by EuroParc and others, including the University of Toronto and Bellcore. The aim of the work at EuroParc was to turn companies into virtual offices. Systems such as OfficeShare, Glance and Portholes allowed employees to see and keep track of each other using in-office video cameras. A fourth system, Background, was designed for meeting spaces to allow employees to monitor the arrival of colleagues in boardrooms.

The Role of GIS

All of the aforementioned memory and intimate technology systems have a strong spatial component. The systems keep track of one or more individuals' activities over time; and, over time, we tend to move through space. This seems to be a golden opportunity to extend the reach of GIS technology into a new area. How much more interesting and powerful would these systems be if combined with GIS? [Update: It is intriguing to note that since this column was originally written the challenge has been taken up by the GIS community and that there are now a series of workshops being held on Ubiquitous Geographical Information Services (see *www.ubigis.org*).]

Mik Lamming, one of McCrone's interviewees and a senior researcher at EuroParc, regretted the limited spatial reach of the presently developed systems, but speculated on taking them outdoors. The obvious way to do this would be to marry EuroParc's research with intelligent vehicle/highway system technology. Again GIS should be part of the picture. In addition, research since the 1970s on "Time Geography" (Haggett et al., 1977, p. 17) should provide fruitful conceptual models and an empirical tradition for analyzing and making use of the data collected by these systems. [Update: This call for a renewed interest in Time Geography and urban accessibility issues has been embraced by numerous geographers and others. Most prominent among these has been Mei-Po Kwan from Ohio State University and her former doctoral student Joe Weber. See, for example, the *Special Issue of the Journal of Geographical Systems*, Volume 5, Number 1, May 2003, edited by Mei-Po Kwan, Donald Janelle and Mike Goodchild.]

Concerns for the Future

Cooke and Burlingame's discussion of the inadequacy of human memory ends with the former exclaiming in distress that "naught is immutable and sure" and the latter declaring that this recognition is "the first step on the road to heaven." If this is so, will our reliance on computerized, intimate technology be our first step in the other direction? McCrone's frequent reference to Orwell's *1984* must surely give us pause for thought. Raising concerns relating to the rise of a surveillant society.

(This column originally appeared in the August 1994 issue of GIS World.*)*

What Can We Learn From Google's Success?

By Daniel Sui

When Pulitzer Prize winning columnist Tom Friedman wrote in the *New York Times* last year "Is Google God?", the response he got was "Is Friedman an Idiot?" (*www.google-watch.org/friedman.html*). Obviously, not everybody was taking Friedman's comments seriously, including those in the GIS community.

But with Google's acquisition of Keyhole—a start-up 3-D mapping company—and the launch of local Google searches with explicit geographic specifications (*local.google.com, maps.google.com, or earth.google.com*), everyone in the GIS industry should take notes (and a deep breath).

The new enhanced Google search capabilities put pressure on MapQuest.com and Yahoo maps as well as pose new challenges and opportunities for the GIS industry. If Keyhole's acquisition is an indication of Google's increasing recognition of the importance of GIS and mapping, it's not too late to ponder what the GIS industry can learn from Google's success and what the long-term implications of Google's encroachment on GIS are.

Why Google?

Google is a play on the word "googol," an arcane mathematical term that refers to the figure of 1 followed by 100 zeros, which reflects the company's ambitious goal to organize the immense amount of information available on the Web.

Google wasn't part of the popular lexicon a few years ago. Now more than 82 million unique users are Googling around the world every month, and the number of Google users is increasing by the minute.

Furthermore, Google seeks to become the gatekeeper for the public Web as well as the "dark" or hidden Web of private databases, dynamically generated pages, controlled-access sites and Web servers within organizations, which was estimated to be tens or even hundreds of times larger than the public Web (Ferguson, 2005).

Google realized that only a small fraction of the vast amount of digital information is indexed. In addition, the proliferation of new computing platforms, especially the increasing cross-pollination of data among new computing devices, will make search engines indispensable.

Rising from the Bottom

With its humble beginnings in a Stanford classroom by two young graduate students in their twenties, Google's success makes a fascinating case for new business models in the digital economy. Many of the concepts central to the high-tech economy (Arthur, 1994) seem applicable to Google's success. It will be interesting to watch how these concepts play out as the story of Google unfolds.

So far, Google has explicitly stated that its strategy isn't a portal strategy. Instead, Google has adopted a "bottom-up" approach, focusing on individual functions that then are integrated into the overall model.

The acquisition of Keyhole and the expansion of Google's geographic search capabilities should be considered good news for the GIS industry for obvious reasons. Google's new geographic capabilities will provide millions of Web users with an enormous amount of geographic information that may have otherwise been missed.

Contrary to some of the dire predictions in the late 1990s, such as the death of distance or irrelevance of geography, Google's expanded search capabilities have proved the opposite one more time: the revenge of geography has taken place. In fact, maps and driving directions were ranked the 5th and 7th most frequently searched words by Google users worldwide last year.

With the maturing of various location-based services on wired and wireless communication networks, it can be anticipated that the demand for geospatial information will continue to grow. Undoubtedly, Google's new geographic capabilities will contribute to make geographic information an integral part of people's daily life. Despite Google's denial, I believe that it has become a portal in general, and a geoportal in particular, besides its status as the "search king."

New Geoportal World

The GIS community has worked for years to build distributed GIS technology that supports geoportal implementation. In metaphorical and literal ways, Google has become a geoportal that few had anticipated. Considering the fact that each entity in the virtual and physical world can and will be georeferenced and tracked with explicit geographic information, the entire world will eventually become a giant database to be searched. How to make sense of this "brave new world" will be the business of search engine companies as well as a top priority for the GIS industry if we want to survive in this competitive environment.

Furthermore, it may be worthwhile to explore to what extent Googling principles can be applied to searches in geographic databases of a smaller scale (e.g., the conventional GIS databases at global, national and local scales). Can the RankPage algorithm, one of the core ideas behind the Google search engine, be adapted to a new data mining technique for GIS users confronted with large geographic databases?

I believe Google's success has ontological and methodological implications for GIScience. Many of the aforementioned issues, ranging from privacy, equity, access and efficiency, have become more acute with Google's capability expanding into the geographic domain.

With the beta version of the MSN search engine equipped with geographic search capabilities, Microsoft also has increasingly recognized the enormous market for search-engine business and geographic information. Now that Microsoft has jumped on the search-engine development bandwagon, it remains to be seen whether Google will suffer the same fate as Netscape did 10 years ago.

Systematic and Regional?

No matter what happens in the next couple years, the booming search business revealed one aspect nobody should ignore in the digital age: equipping users with easy-to-use search tools to find information they need is as important (if not more important) as creating the information itself. Those unable to link to a popular search engine may not be able to disseminate the information they produce to a larger audience, regardless of the information's quality. This may be another source that contributes to the digital divide.

In many significant ways, new technological innovations may have made us better appreciate the dual importance of systematic and regional geography. Those familiar with the history of geographic thought can recall the (sometimes) acrimonious debates among geographers on whether geography should focus on systematic (study one theme intensely) or regional (study multiple themes holistically for a single region) approaches.

By analogy, the portal development strategy takes a "regional" approach in the sense that each portal aims to provide users a gateway/entry point to a variety of information and services, whereas Google and the other Web search engines essentially play the role of "systematic geography" by organizing a particular theme of information in some meaningful way.

Just like geography needs a combination of systematic and regional approaches to flourish, we also need portals and search capabilities to survive in today's digital world. Indeed, I attribute Google's success so far to this hybrid strategy.

By posing the rhetorical question "Is Google God?", Friedman aimed to capture the public's attention on the crucial importance of Google in our lives. All conscious human beings are on a constant search during their lives, although a large number of them don't necessarily know exactly what they're searching for. So part of Google's success is perhaps due to the fact that it has met one of our higher spiritual needs.

Sergey Brin, co-founder with Larry Page of Google Inc., once remarked that "the perfect search engine would be like the mind of God." How interesting (and inspiring) to observe that the ultimate goal of developing a search engine has become the same quest as Stephen Hawking's pursuit in theoretical physics: to understand "the mind of God."

Although it's premature to conclude whether it's the grand unified theory in physics or a perfect search engine that should represent the ultimate triumph of human reason, one thing for sure is that both endeavors have made us contemplate further on what "G" should truly stand for.

(This column originally appeared in the March 2005 issue of GeoWorld.*)*

Will GIS Last?

By Nigel Waters

We know GIS is here to stay. GIS is the technology for the 21st century. But what about individual GIS records and databases? Will they still be usable 100 years from now? Recent commentaries provide reasons for pessimism and optimism (Peuquet, 1994; Campbell, 1995; Rothenberg, 1995).

The Impermanency of Digital Records

A paper map is a spatial database that affords a high degree of permanence and accessibility. A paper map is permanent, because nothing short of fire or water is likely to damage it irreparably. And it is accessible, because it usually has an accompanying legend that explains the symbology. Besides fire and water, the only danger to paper maps is that people get tired of storing them.

Let me illustrate the relative advantages of the permanency of paper and digital spatial databases with a story from my past. When I was exploring the rural farm settlement around the city of Regina in Saskatchewan, Canada, in 1971, it broke my heart to learn that every time the Saskatchewan Municipal Road Assistance Authority produced a new map the old ones were destroyed. Thus, paper maps do have their problems, but as Campbell (1995) points out, copyright libraries traditionally have stored one record of each map printed.

Now consider what happened when I located all the farmsteads in my Saskatchewan study areas. I recorded the farmstead locations in Cartesian coordinates on the respective township maps and then returned to Cambridge University and asked if I could use the computer to analyze my data. My supervisor asked me in turn, "Why on Earth would a geographer ever want to use a computer?" Eventually I convinced him that this was a good idea, and I entered my data into the computer on paper tape.

A year later when I arrived at the University of Western Ontario for graduate studies, I discovered to my alarm that paper tape was passé. So the paper tapes were translated with considerable difficulty to the then state-of-the-art punched cards.

I carried my new database of punched cards with me to the University of Calgary in 1975. But it was only a few years later that we were informed that the card readers were being removed, and we had better read our punched cards onto seven-track magnetic tape. I remember the heartbreak as decks of two thousand cards would be chewed to pieces by the high-speed card readers and would have to be repunched. Then we changed to nine-track tapes. Then we changed to computing systems and wanted our data on 5.25-inch disks, and then on 3.5-inch disks, and then on optical disks and hard drives formatted this way and then formatted that way. Indeed, it was only a few months ago that I was asked for the data on the fire-alarm-demand surface from my doctoral thesis (see Lecture 28 from the National Center for Geographic Information and Analysis), and to my amazement I realized the only copy I had was on punched cards.

OK, so it was a long story, but it shows that Rothenberg (1995) is right in his concern that digital media have a short life, and the rapid advances made almost annually with respect to storage media and software formatting render documents unreadable.

Obsolescence vs. Physical Deterioration

The real danger is not the impossibility of reading old GIS databases (although that eventually becomes a problem if no one has the right hardware or software), but it is the cost of reading the database if the hardware and/or software are simply hard to find. Cost, after all, is why the authorities in Saskatchewan were not keeping a historical record of their settlement maps. Cost is why the U.S. Coast and Geodetic Survey dropped all reference to seismic activity between 1840 and 1961 in producing a map of plate tectonics (Weizenbaum, 1984, p. 237). The post-1961 data were in machine-readable form, the earlier data were not. And cost is why the British Ordnance Survey has managed to avoid any legal requirements to deposit digital data, thus jeopardizing the historical record after 1997 when the present microfilming obligations end (Campbell, 1995).

The second problem is not obsolescence, but the physical life of the media—the lifetime over which we can expect no deterioration, no loss of data. According to Rothenberg (1995, p. 44), sometimes obsolescence occurs before expected physical deterioration, and sometimes it is the other way around. He asserts that magnetic tape can be expected to be obsolete in five years, but its physical lifetime is only one year. Optical disks, on the other hand, may take 10 years to become obsolete but their physical lifetime, without any loss of data, is expected to be 30 years. It can be much worse than that. The 21MB floptical I bought last year appears, like betamax video machines, to have become instantly obsolete.

A Source of Optimism

Peuquet (1994) provides a source of optimism: the creation of GIS databases that incorporate a temporal component. Thus, the changing geography of a country would be incorporated as part of the temporal attributes within its proposed triadic framework for a temporally based GIS. Conceptual frameworks, such as Peuquet's, which permit the development of temporal GIS, may be the only hope for the future. It is unlikely that Peuquet has resolved all the problems facing the development of temporal GIS, and it is equally unlikely that a single solution will suit the myriad of applications for which GIS are relevant, but Peuquet's is a step in the right direction. Without the development of temporal GIS, Weizenbaum's (1984) comment, "The computer has thus begun to be an instrument for the destruction of history (page 238)," may appear to be all too prescient.

(This column originally appeared in the March 1995 issue of GIS World.*)*

Geographer Questioned at Pearly Gates

By Jerry Dobson

Based on a lifetime of personal experience, I fully expect the following conversation when my time comes to approach the Pearly Gates:

St. Peter says, "So, you're a geologist."

"Uh, no sir, that's 'geographer.' I'm a geographer."

He flushes. "Well, it's an easy mistake to make. Not much difference is there?"

Not wanting to cause further embarrassment at this exceedingly inopportune moment, I say in my most conciliatory way, "Right, sir. They do have four letters in common, and three of them are right up front. Scores of universities—even the National Academy of Sciences—have made the same mistake. But, clearly, I'm a geographer."

"Ah, a mapmaker. Now there's a noble profession. When I was a boy, I used to ..."

"Pardon me, sir, but you're thinking of 'cartographer.' I'm a geographer. We've been around for 2,500 years or so. Surely you've met Strabo, Ptolemy, Herodotus ..."

"Before my time, but I've heard of them. Several Ptolemys, in fact ... pharaohs, I believe, but they didn't make it in. And, oh yes, later there was an astronomer ... same name—not a pharaoh."

"Well, yes, astronomer, but also an accomplished mathematical cartographer and geographer. He invented latitude and longitude," I say.

"But this Herodotus chap ... Father of History, wasn't he?"

"His works are called *The Histories*, but back then 'history' just meant an investigation of any sort, and his investigations most certainly were geographical. History, as we know it today, wasn't invented until hundreds of years later. Make no mistake. Ptolemy, Herodotus and I are geographers!"

"Right, right. So where did you teach?"

"I taught many years at the University of Kansas, but don't assume all geographers are teachers. In fact, I myself worked at Oak Ridge National Laboratory for 26 years before moving to academia."

"What on Earth might a geographer do at a national laboratory?"

"Oh, facility siting, transportation management, logistics, environmental analysis, GIS ..."

"GIS! What's that?"

"That's the computerized form of geography. It's the hottest thing on Earth. I'm surprised you haven't heard of it."

He flushes again, and I realize I've gone too far. He stares down the never-ending line behind me. "Anybody back there heard of something called GIS?"

An eager plaintiff shouts back, "It's a little gizmo that collects satellite signals and tells you exactly where you are."

Quickly, another one shouts, "You doofus, that's GPS. GIS is a computer map." His attitude doesn't serve him well, and he drops (literally) out of line.

Yet another calls more humbly, "It stands for 'graphic information system.' We use them all the time where I work ... er, worked."

"No, it's a *geo*graphic information system," I say, "and it's a lot more than a computer map. It handles attributes, geometry, topology ..."

"Ah, topology," St. Peter says. "Why didn't you say so? Hills and valleys I understand."

"No, sir. You're thinking of topography. Topology is how we represent spatial relationships. Today, many disciplines are working together to develop new ontologies of ..."

"Ontologies? Got that from biology, didn't you? 'Ontology recapitulates something or other,' as I recall."

"No, sir. You're thinking of ontogeny. Biologists used to say 'ontogeny recapitulates philogeny.'"

"Enough, enough," he whines. "What's it good for anyway, this GIS?"

"All the things I mentioned earlier, plus a lot more. I've used it myself to study lake acidification, human evolution ..." He frowns, and I quickly move on, "continental drift ..."

"See, now *that's* geology," he says.

"No, sir. Even geologists call that field paleogeography. As a matter of fact, continental drift was first proposed by the geographer Ortellius in 1596, based on the fit of coasts between Africa and South America, and geologists didn't catch on for another 366 years."

"I noticed it myself sometime around 800," St. Peter says, "but I didn't say anything." Then, he whispers, "They actually fit better if you rotate South America 90 degrees. And there are some other fits around the world that you just wouldn't believe." Recovering, he adds, "Great view from up here, of course, but it must have been difficult for your Ortellius with nothing more than crude, large-scale maps."

"Actually we call them small-scale maps when they show the whole globe," I say. Again St. Peter flares, and I try to make amends, "But a lot of otherwise intelligent mortals call them large scale. I've witnessed whole conversations proceed from beginning to end with people using the terms in opposite ways and yet never knowing. It's so confusing, I've stopped using the terms myself."

"Anyway, so we gave you satellites, and now you can map land use to your heart's content from hundreds of miles up."

"You mean land cover," I say. "Land cover is physical, tangible and visible even from satellites. Land use is economic and cultural. Sometimes we can infer use from cover, but not always. Often the only way is to knock on doors and ask people how they use their land."

"Back to the business at hand. Have you done anything important?" he asks. "I've got to put something down."

I ponder a moment. "I was president of the American Geographical Society."

"Wow!" he says. "Nice magazine—great videos, your own TV channel and now movies. Impressive."

"You're thinking of the National Geographic Society," I say. "We're older and more ..."

"Enough of this chitchat," he says. " I'm afraid you geographers just don't fit in up here. It's a peaceable place, and you're so contrary."

"We'll get along," I say. "We spend our lives accommodating folks who don't understand us any better than you do. Still, we aren't much appreciated by the so-called 'hard' sciences. Is that going to be a problem?"

"Not really," he says. "We don't have all that many. But watch this ..." He turns to the courtyard behind him. "Engineers!" he calls, and wings flutter all around.

I can't hide my shock.

"Oh, yes," he says. "Quite a few make it." Then in a lower voice, "Simple lot, but they're good with numbers. We use them for counting sparrows and tallying things. And some of them are good with a harp."

"We're good with engineers," I say.

"Still," he insists, "it would help if you had some redeeming quality."

I search the deepest, darkest recesses of my soul. "I was a chemistry major my first two quarters of college," I admit.

"Enter," he says.

I dash in and quickly mingle into the crowd, pretending not to hear as he returns to his ledger and mutters, "Now, how is it you spell 'spatial?'"

(This column originally appeared in the September 2002 issue of GeoWorld.*)*

Section 7.
References

Ames, A.L, Nadeau, D.R., Moreland, J.L. 1997. *The VRML 2.0 Sourcebook* (2nd Edition), John Wiley & Sons, New York.

Arthur, W.B. 1994. *Increasing Returns and Path Dependence in the Economy.* University of Michigan Press, Ann Arbor, Mich.

Barth, J. 1960. *The Sot-Weed Factor*. Doubleday, New York (Reference is to the 1969 Bantam edition).

Bertazzon, S., Waters, N.M. 1996. "The Use of GIS Applications on the Internet for Marketing Tourist Destinations." *Proceedings of GIS/LIS '96*, American Congress on Surveying and Mapping (ACSM)/American Society of Photogrammetry and Remote Sensing (ASPRS), Bethesda, Md., pp. 870-881.

Campbell, T. 1995. "The Digital Downside."" *Geographical*, Vol. 66, No. 12, p. 17.

Dobson, J.E., 1999. "Explore Aquaterra Lost Land Beneath the Sea." *GeoWorld* Vol. 12, No. 1, p. 21.

Fairbanks, R.G. 1989. "A 17,000-year Glacio-eustatic Sea Level Record: Influence of Glacial Melting Rates on the Younger Dryas Event and Deep-ocean Circulation." *Nature,* Vol. 342, pp. 637-642.

Ferguson, C.H. 2005. "What's Next for Google?" *Technology Review*. Available online at: *www.technologyreview.com/articles/05/01/issue/ferguson0105.asp?p=1.*

Haggett, P., Cliff, A. D., Frey, A. 1977. *Locational Analysis in Human Geography*. Arnold, London.

Hertzberg, H. 1999. "The National Geographic Society Goes Where Many Have Gone Before." *The New Yorker*, August 23 & 30, 1999, p. 62.

Kemp, K. 1996. "The New On-Line Core Curriculum in Geographic Information Science." *Proceedings of GIS/LIS '96* , ACSM/ASPRS, Bethesda, Md., pp. 363-367.

Mander, J. 1992. *In the Absence of the Sacred: The Failure of Technology and the Survival of the Indian Nations*. Sierra Club Books, San Francisco.

McCrone, J. 1993. "Computers that Listen." *New Scientist*, Vol. 140, No. 1902, pp. 30-35.

McCrone, J. 1994. "Don't Forget Your Memory Aide." *New Scientist*, Vol. 141, No. 1191, pp. 32-36.

Norman, D.A. 1988. *The Design of Everyday Things*. Doubleday/Currency, New York.

Norman, D.A. 1993. *Things That Make Us Smart*. Addison-Wesley, Reading, Mass.

Peltier, W.R., 1994. "Ice Age Paleotopography." *Science,* Vol. 265, pp. 195-201.

Pequet, D. 1994. "It's About Time: A Conceptual Framework for the Representation of Temporal Dynamics in Geographic Information Systems." *Annals of the Association of American Geographers*, Vol. 84, pp. 441-461.

The Professional Geographer. 1999. "Focus: Geography of Ocean-Space." Vol. 51, No. 3.

Rothenberg, J. 1995. "Ensuring the Longevity of Digital Documents." *Scientific American*, Vol. 272, No. 1, pp. 42-47.

Skupin, A., Buttenfield, B.P. 1996. "Spatial Metaphors for Visualizing, Very Large Data Archives." *Proceedings of GIS/LIS '96*, American Congress on Surveying and Mapping/American Society for Photogrammetry and Remote Sensing, Bethesda, Md., pp. 607-17.

Spackman, K., Tan, S. 1993. "When Turning Gets Tough." *New Scientist*, Vol. 137, No. 1864, pp. 26-31.

Stephenson, N. 1993. *Snow Crash*, Bantam Books, New York.

Stringer, C., Gamble, C. 1993. *In Search of the Neanderthals: Solving the Puzzle of Human Origins*. Thames and Hudson, New York.

Verne, J., 1863. *Five Weeks in A Balloon.* Ameron Limited, Mattituck, New York.

Want, R., Hopper, A., Falcao, V., Gibbons, J., "The Active Badge Location System." *ACM Transactions on Information Systems*, Vol. 10, No. 1, pp. 91-102.

Weizenbaum, J. 1984. *Computer Power and Human Reason* (2nd edition) Penguin, Harmondsworth, England.

Wright, J. K. 1952. *Geography in the Making; The American Geographical Society1851-1951*, The American Geographical Society, New York.

Wurman, R.S. 1989. *Information Anxiety*. Doubleday, New York.

Afterword—And Now to Bed

By Mike Goodchild

Scholarly writing is notorious for its pedantry, for using words of five syllables like methodology or terminology when shorter words like method or term would do, and for devoting entire volumes to abstract questions not far removed from "How many angels could dance on the head of a pin?" It is said that the typical academic paper, published after long and exhaustive review in the pages of an expensive journal, is eventually read by no more than a handful of people, including the editor and reviewers. Academic writing has evolved its own dry style that is stripped of the kinds of devices used by lecturers to keep students awake—wit, hyperbole, graphics—resulting in what may well be the ultimate soporific.

Occasionally, however, academics allow themselves the luxury of writing—and reading—work that is brief, sparkling, and entertaining. We often think of such luxuries as associated with particular locations, such as the beach, the armchair, or the bed. We may well learn more from the wide range of ideas in a good, well-edited *New Yorker* article than from an entire volume of our favorite learned journal, and may find the entertaining style of Eric Newby, Paul Theroux, or Bruce Chatwin far more insightful in what it reveals about the geography of remote parts of the world than any amount of more scholarly writing. What geographer has not enjoyed Dava Sobel's (1996) account of the measurement of longitude, or Simon Winchester's (2002) books on geologic mapping, Krakatoa (2003), or the Yangtse (2004)? My own current light reading includes *Mapping Hacks: Tips and Tools for Electronic Cartography* (Erle et al., 2005), a book that contains not a single reference to the academic cartographic literature.

How interesting, then, to be asked to write an afterword to an entire book of short, entertaining pieces on matters of fundamental significance to anyone interested in geography or geographic information systems. All of them are stimulating, raising important questions in a style that is designed to invite rather than close off further thinking. Some reflect our deepest, darkest concerns: "Have I picked a field of interest that is fashionable today, but destined for the intellectual wastebasket in a few years; or will I be feted twenty years from now as the leader who saw the potential of a new kind of science?" Others catch the excitement of the day over such new developments as Digital Earth or location-based services.

Academics write such things at their peril, of course, and the articles that have been assembled in this volume are beyond the academic pale—unrefereed, short, and without the credentials necessary to count for tenure, promotion, or honors. It is only a few years since a certain popularizer of science whose name is known to millions was rejected for membership by a floor vote of the National Academy of Sciences, despite having a sterling academic reputation. We live in an academic world in which terms like useful and popular are sometimes so unusual as to seem at times pejorative. So we should recognize that the pieces in this book were not written in the interests of career advancement, financial reward, or academic reputation but because the authors felt that their subjects needed airing before a wide audience that shared their interest. We forget how much of intellectual life is driven by such altruism—why else would an academic abandon home and family to spend tens of hours crammed into the economy section of an airplane in order to attend a conference half a world away?

The pieces in the book are written by North Americans, and they make most sense to a North American audience. In other parts of the world geography is a very different discipline. In the United Kingdom, for example, it is one of the most popular high-school subjects, although no longer compulsory, and a strong geography program such as at the University of Leeds can expect six or more applicants for every one of the 200 places it offers annually in its undergraduate program. Geography in the United States is a comparatively small, anxious discipline that must continually fight for students, most of whom will have arrived on campus with almost no awareness that the discipline even exists. No wonder, then, that North American academics can write about GIS saving geography.

But why, at the end of the day, should anyone write about the relationship between a software package and a discipline? It must seem very odd to historians or ecologists that geographers write so much about GIS, and odd that anyone would devote an entire book to "geospatial matters", or think that a fairly straightforward computer application could spawn an entire new discipline of geographic information science. I can't imagine anyone writing about Microsoft Word saving the Modern Language Association, or asking whether Microsoft Excel can prevent war, though some fairly profound things have been said about Microsoft PowerPoint by the likes of Edward Tufte. GIS and geospatial tools are certainly complex, and building them is a challenge, but weren't the problems essentially solved many years ago, in the early days of GIS and remote sensing? Before you put the book back on the bedside table and turn out the light let me make two suggestions.

First, geography in the United States is a small, weak discipline, yet one with a clear view of its potential significance as a field that straddles and links the social and environmental sciences, and as a field with a claim to know more than any other discipline about the nature and importance of space. But this is a vast territory, and a very significant slice of the intellectual pie. Covering it would require, as Richard Atkinson once pointed out when he was President of the University of California, a department far larger than departments are expected to be, and a share of the university's resources far larger than any department is allowed to have—it would require a victory in the interdepartmental wars that would be more overwhelming than Henry V's at Agincourt or Washington's at Yorktown. In this context GIS provides a way of extending and enlarging geography's role, of delivering spatial perspectives and tools across the academic landscape to an extent that far exceeds the capacity of geography itself. GIS is the academic mouse that roared.

Second, spatial really is special, and far more significant than the conceptual basis of Word or Excel. Luc Anselin (1989) was perhaps the first to point out that spatial is special in two distinct and significant ways. What he called spatial heterogeneity, or the fact that the Earth's surface exhibits what a statistician would call non-stationarity, flies in the face of a scientific tradition that insists that no result is useful unless it applies everywhere at all times. Geographers have struggled to find laws of geography that could rank with Newton's Laws of Motion, and have largely abandoned the search. But does this mean that geography can never rank with physics as a respected discipline, because it cannot find anything to say about the Earth's surface that is similarly general? No, in this context the recent growth of interest in GIS reflects a fundamental departure, towards a new kind of scientific thinking in which the variation of things from place to place is of primary importance, and in which computers operating on massive amounts of data provide the only way to extract useful results. An entire artillery of new methods of spatial analysis has been developed in the past decade or so under such headings as place-based or local analysis, and these methods are inseparable from the GIS that make them practical.

Anselin's second characteristic was spatial dependence, and here again we find that trying to apply traditional scientific thinking to space is fraught. The methods of inferential statistics developed by R.A. Fisher and others were based on controlled experiments, such as fields planted with different crops, where it is reasonable to assume that each sample in the experiment is essentially independent of each other sample. But Tobler's First Law states that exactly the opposite is normally the case in space; that the assumption of independence is untenable. Again what appears to be no more than a simple tool for analyzing maps turns out to represent a profound departure from the traditional ways we humans have approached scientific questions.

So GIS really is transformative, in the sense that the term is used in discussions of science policy, and the pieces in this book are a brief but illuminating set of variations on that basic theme. Whether this point will leave you contentedly reassured that you made the right choice in pursuing geography and GIS, and send you blissfully to sleep, or whether it will keep you awake, tossing and turning and cursing the hand that wrote it, only time will tell.

References

Anselin, L. 1989. "What's Special about Spatial Data: Alternative Perspectives Spatial Data Analysis." *NCGIA Report 89-4*, Santa Barbara, Calif.

Erle, S., Gibson, R. and Walsh, J. 2005. *Mapping Hacks: Tips and Tools for Electronic Cartography*. O'Reilly Media, Sebastopol, Calif.

Sobel, D., 1995. *Longitude: The True Story of a Lone Genius Who Solved the Greatest Scientific Problem of His Time*. Walker, New York.

Winchester, S., 2001. *The Map that Changed the World: William Smith and the Birth of Modern Geology*. HarperCollins, New York.

Winchester, S., 2003. *Krakatoa: The Day the World Exploded, August 27, 1883*. Harper-Collins, New York.

Winchester, S., 2004. *The River at the Center of the World : A Journey up the Yangtze, and Back in Chinese Time.* Picador, London.